Simple**R**
Using Machine Learning Algorithms in R

Darrin Thomas

Simple**R**
Using Machine Learning Algorithms in R

Darrin Thomas
Asia-Pacific International University

SuJinSoLa
Saraburi, Thailand

Layout: Darrin Thomas
Photo Researcher: Darrin Thomas
Cover Design: Darrin Thomas

SuJinSoLa

ISBN-13: 978-1546324195
ISBN-10: 1546324194

About the Author

Darrin Thomas, PhD, grew up in Sacramento, California and has over ten years of experience as  a teacher and lecturer from Kindergarten to graduate school. He completed his bachelor and master degree in saxophone performance at California State University Sacramento. After working as a substitute teacher, he completed a credential in teaching at Pacific Union College. He then worked as a music teacher before moving to Thailand to work as a lecturer in Education/Psychology Department at Asia-Pacific International University (APIU). While overseas, Dr. Darrin completed his master degree in education at APIU. He then moved to the Philippines and completed his doctoral degree in education at Adventist International Institute of Advanced Studies. Currently, Dr. Darrin is a lecturer at Asia-Pacific International University. His enthusiasm for machine learning has led to works involving many different algorithms applied in an educational context.

DEDICATION

To my wife and children

Table of Contents

About the Author iii

Preface x

Chapter 1 Multiple Regression 1

Chapter 2 Best Subset Regression 14

Chapter 3 Ridge Regression 27

Chapter 4 Lasso Regression 36

Chapter 5 Elastic Net Regression 43

Chapter 6 Linear Discriminant Analysis 51

Chapter 7 Quadratic Discriminant Analysis 60

Chapter 8 Logistic Regression 66

Chapter 9 Assessing Classification Model Performance 72

Chapter 10 Assessing Numeric Model Performance 88

Detail Table of Contents

Chapter 1 Multiple Regression 1
    Chapter Objectives 1
    Data Preparation 1
    Missing Data 3
    Normality of Variables 5
    Collinearity...11
    Model Results..13
    Conclusion 13
Chapter 2 Best Subset Regression 14
    Chapter Objectives 14
    Initial Model Development 15
    Subset Regression Model 16
    Selected Variables 18
    Final Model 20
    Conclusion 26
Chapter 3 Ridge Regression..27
    Chapter Objectives 27
    Regularization 27
    Ridge Regression...28
    Data Preparation 28
    Model Development 30
    Model Testing..34
    Conclusion 35
Chapter 4 Lasso Regression 36
    Chapter Objectives 36
    Data Preparation 36
    Model Development 38
    Model Testing 41
    Conclusion 42
Chapter 5 Elastic Net Regression 43
    Chapter Objectives...43
    Data Preparation 43
    Model Development 45
    Model Testing 47
    Cross Validation 48
    Conclusion 50
Chapter 6 Linear Discriminant Analysis..51
    Chapter Objectives 51
    Data Preparation 51
    Model Development...57
    Model Testing 58
    Conclusion 59
Chapter 7 Quadratic Discriminant Analysis 60
    Chapter Objectives 60

Data Preparation 60
Model Development (LDA)..61
Model Testing (LDA) 63
Quadratic Model Development and Testing..64
Conclusion 65
Chapter 8 Logistic Regression 66
Chapter Objectives...66
Key Terms Related to Logistic Regression 66
Data Preparation 67
Model Development..69
Model Testing 71
Conclusion...71
Chapter 9 Assessing Classification Model Performance..........................72
Chapter Objectives...72
Confusion Matrix 72
Assessing Models with Confusion Matrices Outputs 74
Confusion Matrix Explanation..77
Using Visuals to Assess Classification Models 79
Cross-Validation..85
Conclusion...86
Chapter 10 Assessing Numeric Model Performance 88
Initial Model Development 88
Second Model..91
Cross-Validation..93
Conclusion...94

Preface

The idea of data science and the search of big data has truly taken the world by storm. Understanding how to analyze data and answer research questions has quickly become more and more important.

My foray into machine learning was in many ways by accident. I was searching for a way to do structural equation modeling without have to spend huge amounts of money only professional software. This search led me to R. While learning R I stumbled upon machine learning, got really excited about all of these free tools at my disposable, and the rest is history.

This text was written to provide a condense resource that deals with many different commonly used machine learning algorithms. The desire was to develop something easy to read while providing practical applications. Short and to the point was the mission in communicating this information.

One assumption of this text is that the readers is already familiar with R and R studio and are able to read and interpret the code in the examples. Having said that, I make no claim to expertise in R but I do know how to get things done when using this programming language. Steps were taken to try to make the text as simple as possible and sympathetic to those who may not have as strong a background in programming.

This book does not provide a thorough explanation of the statistical mechanics of the various algorithms. This is a "how to" text not a "why to" text. A dense statistical explanation would have deadened the text as well as lengthen with information that the more practical-minded may not have appreciated. Therefore, this assumes that the reader knows what they want to do by lacks the tools to achieve it. There is no explanation of data science or machine learning as these are concepts that one should be familiar with when examining the ideas if this text

Each chapter 1-8 each deals with the application of a specific algorithm. The thrust of the book is the use of regression and feature selection for regression, which is covered in chapters 2,3,4, and 5. The only departure from this theme is when we examine categorical models in chapters 6,7, and 8. Chapters 9-10 deal with how to assess classification models and numeric models respectively.

Chapter One: Multiple Regression

Multiple regression is the workhorse of machine learning. It works in most instances and in many ways, most of the other linear algorithms are variations on multiple regression. Due to the importance of multiple regression, it is necessary to review the assumptions of regression. This is done in the current chapter, as I did not want to waste space documenting the assumption check in every chapter of the book.

There are several assumptions of multiple regression. The ones we will address are as follows

- Normality of variables
- Linear Relationship between the outcome variable and the independent variables.
- Presence of outliers
- No Multicollinearity
- Homoscedasticity

Another concept that is not an assumption but needs to be address is missing data. We will cover this as well.

The objectives of this chapter are as follows...

Chapter Objectives

- Prepare data for analysis
- Demonstrate how to deal with missing data
- Develop a multiple regression model
- Address the assumptions of multiple regression

Data Preparation

We will attempt to explain the salary of a baseball based on several variables. The first thing we need to do is load our data. Our "Hitters" data comes from the "ISLR" package and we will use the data set "Hitters". There are 20 variables in the dataset as shown by the "str" function.

```
library(ISLR);library(corrplot);library(mice);library(lmtest)
data("Hitters")
str(Hitters)
```

```
## 'data.frame':    322 obs. of  20 variables:
## $ AtBat    : int  293 315 479 496 321 594 185 298 323 401 ...
## $ Hits     : int  66 81 130 141 87 169 37 73 81 92 ...
## $ HmRun    : int  1 7 18 20 10 4 1 0 6 17 ...
## $ Runs     : int  30 24 66 65 39 74 23 24 26 49 ...
## $ RBI      : int  29 38 72 78 42 51 8 24 32 66 ...
## $ Walks    : int  14 39 76 37 30 35 21 7 8 65 ...
## $ Years    : int  1 14 3 11 2 11 2 3 2 13 ...
## $ CAtBat   : int  293 3449 1624 5628 396 4408 214 509 341 5206 ...
## $ CHits    : int  66 835 457 1575 101 1133 42 108 86 1332 ...
## $ CHmRun   : int  1 69 63 225 12 19 1 0 6 253 ...
## $ CRuns    : int  30 321 224 828 48 501 30 41 32 784 ...
## $ CRBI     : int  29 414 266 838 46 336 9 37 34 890 ...
## $ CWalks   : int  14 375 263 354 33 194 24 12 8 866 ...
## $ League   : Factor w/ 2 levels "A","N": 1 2 1 2 2 1 2 1 2 1 ...
## $ Division : Factor w/ 2 levels "E","W": 1 2 2 1 1 2 1 2 2 1 ...
## $ PutOuts  : int  446 632 880 200 805 282 76 121 143 0 ...
## $ Assists  : int  33 43 82 11 40 421 127 283 290 0 ...
## $ Errors   : int  20 10 14 3 4 25 7 9 19 0 ...
## $ Salary   : num  NA 475 480 500 91.5 750 70 100 75 1100 ...
## $ NewLeague: Factor w/ 2 levels "A","N": 1 2 1 2 2 1 1 1 2 1 ...
```

Missing Data

We now need to assess the amount of missing data. This is important because missing data can cause major problems with different analysis. Although it is not a concern for multiple regression it is still good practice to address this.

We are going to create a simple function that will explain to us the amount of missing data for each variable in the "Hitters" dataset. A function is simply a piece of code that automates a task. After using the function we need to use the "apply" function to display the results according to the amount of data missing by column and then by row.

```
Missing_Data <- function(x){sum(is.na(x))/length(x)*100}
apply(Hitters,2,Missing_Data)
##     AtBat      Hits     HmRun      Runs       RBI     Walks     Years
##   0.00000   0.00000   0.00000   0.00000   0.00000   0.00000   0.00000
##    CAtBat     CHits    CHmRun     CRuns      CRBI    CWalks    League
##   0.00000   0.00000   0.00000   0.00000   0.00000   0.00000   0.00000
##  Division   PutOuts   Assists    Errors    Salary NewLeague
##   0.00000   0.00000   0.00000   0.00000  18.32298   0.00000
apply(Hitters,1,Missing_Data) #results are not displayed
```

For each column or variable, we can see that the missing data is all in the salary variable, which is missing 18% of its data. By row (not displayed here) you can see that a row might be missing anywhere from 0-5% of its data. The 5% is from the fact that there are 20 variables and there is only missing data in the salary variable. Therefore 1/20 = 5% missing data for a row.

To deal with the missing data, we will us the 'mice' package. The code is a repeat of our function but using the function 'md.pattern' from the 'mice' package.

```
md.pattern(Hitters)
##      AtBat Hits HmRun Runs RBI Walks Years CAtBat CHits CHmRun CRuns CRBI
## 263     1    1     1    1   1     1     1      1     1      1     1    1
##  59     1    1     1    1   1     1     1      1     1      1     1    1
##         0    0     0    0   0     0     0      0     0      0     0    0
##      CWalks League Division PutOuts Assists Errors NewLeague Salary
## 263      1      1        1       1       1      1         1      1 0
##  59      1      1        1       1       1      1         1      0 1
##          0      0        0       0       0      0         0     59 59
```

This output tells us that 263 observations are complete while 59 are missing data (59 / 263 + 59 = .18). All the missing data is in salary, as we already knew. The code below allows us to impute the missing data. For the details on the arguments inside the "mice" function you can type "?mice " to learn more.

```
Hitters1 <- mice(Hitters,m=5,maxit=50,meth='pmm',seed=500)
```

In the code above, we did the following

We made a new variable called 'Hitters1' and ran the 'mice' function on it. This function made 5 datasets (m = 5) or guesses as to what the salary of the player was and used predictive meaning matching (method = 'pmm') to guess the missing data point for each row (method = 'pmm'). You can view the guesses for each row by the name of the baseball player in the code below. Only the first ten rows are displayed here.

```
Hitters1$imp$Salary[1:10,]
##                      1        2        3        4        5
## -Andy Allanson    75.000  350.000  190.000  100.000  286.667
## -Billy Beane     230.000  185.000   68.000  100.000  100.000
## -Bruce Bochte    612.500  530.000  850.000  215.000  925.000
## -Bob Boone       110.000  400.000  525.000  416.667 2127.333
## -Bobby Grich     320.000  560.000  210.000  780.000  640.000
## -Bob Horner      933.333 1925.571  750.000 1450.000  933.333
## -Bobby Meacham   490.000  155.000  650.000   90.000   70.000
## -Ben Oglivie     420.000 1237.500 1300.000 1008.333 1050.000
## -Bip Roberts      75.000  230.000   70.000  512.500   97.500
## -Bill Russell    650.000  350.000   67.500  400.000  700.000
```

As you can see, each of the datasets vary wildly in terms of salary. One way to determine which to use is to compare the value with the actual values of the examples that are not incomplete. However, for brevity, we will use dataset 1 of the imputed data. Below is the code for using dataset 1.

```
completedData <- complete(Hitters1,1)
```

Normality of Variables

Now we need to deal with the normality of each variable. To save time, I will only explain how I dealt with the non-normal variables. The two variables that were non-normal were "salary" and "Years". To fix these two variables I did a log transformation of the data. This is not the only options as there are several other ways to transform data. However, this was the best option in terms of helping to improve the normality of the data.

The new variables are called 'log_Salary' and "log_Years". Below is the code for this with the before and after histograms. Figure 1.1 and 1.2 are salary before and after transformation. Figure 1.3 and 1.4 are years before and after transformation.

```
hist(completedData$Salary)
```

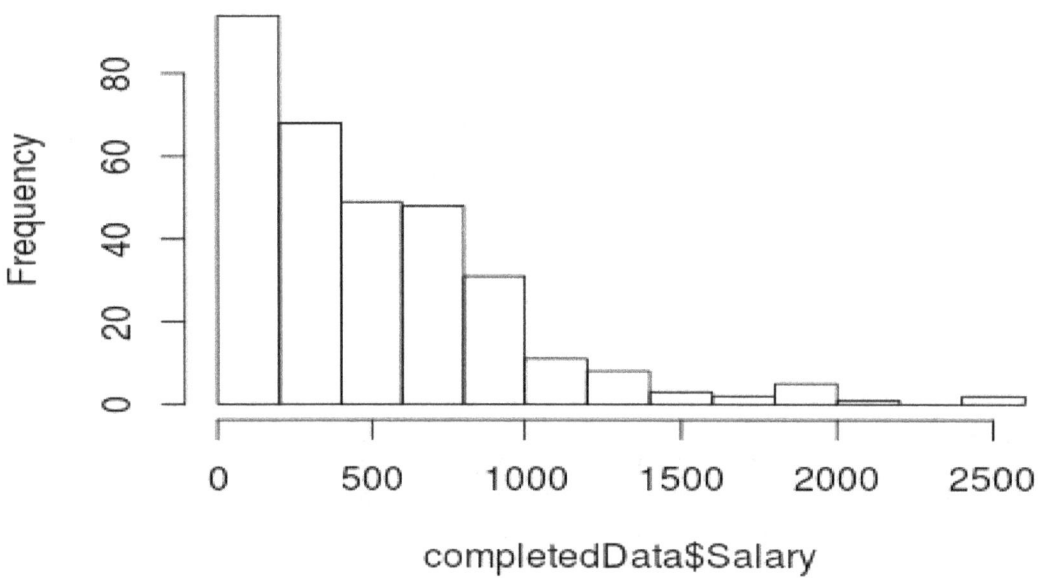

Figure 1.1: Salary before transformation

```
completedData$log_Salary<-log(completedData$Salary)
hist(completedData$log_Salary)
```

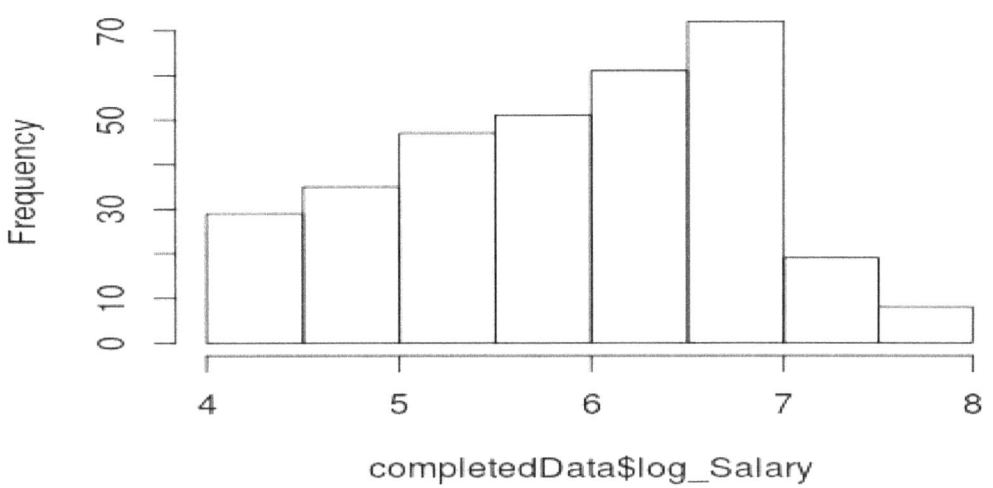

Figure 1.2: Salary after transformation

```
hist(completedData$Years)
```

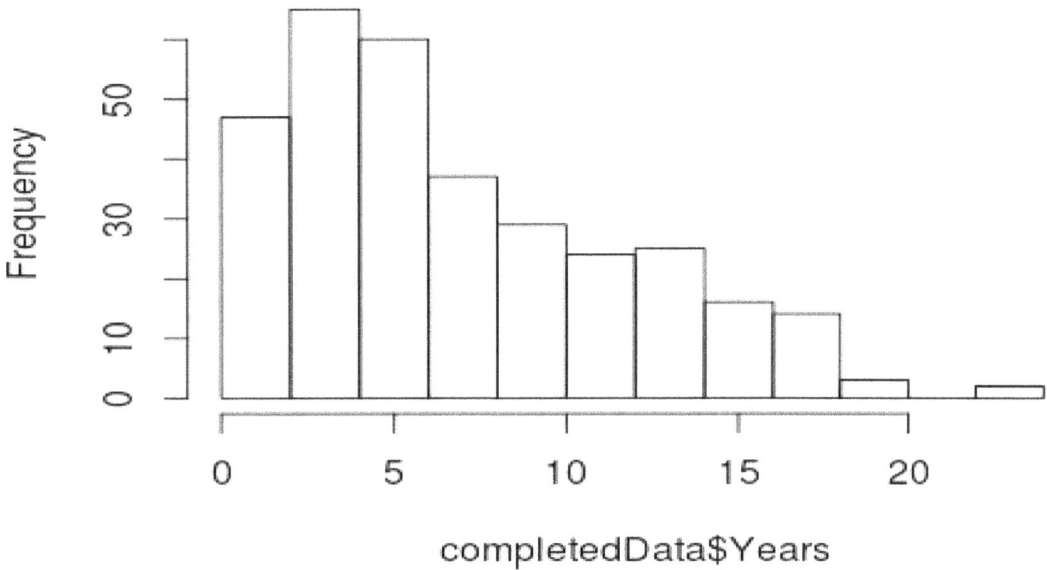

Figure 1.3: Years before transformation

```
completedData$log_Years<-log(completedData$Years)
hist(completedData$log_Years)
```

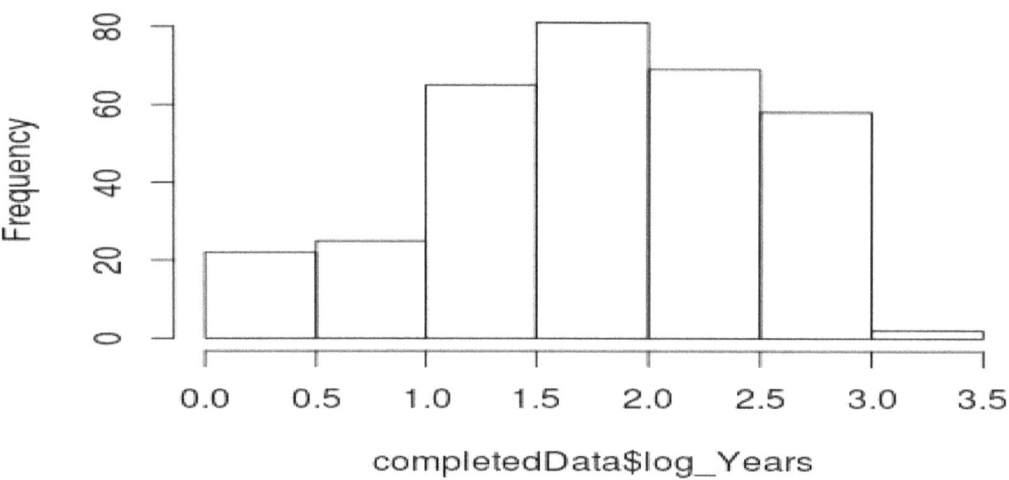

Figure 1.4: Years after transformation

By doing the actual regression analysis, we can also check for outliers as well as the assumption of homoscedasticity and linearity. This is done by using the "plot" function on the created model. Below is the code. The first figure (1.5) is the residuals vs. fitted plot.

```
Salary_Model<-lm(log_Salary~Hits+HmRun+Walks+log_Years+League,
data=completedData)
plot(Salary_Model)
```

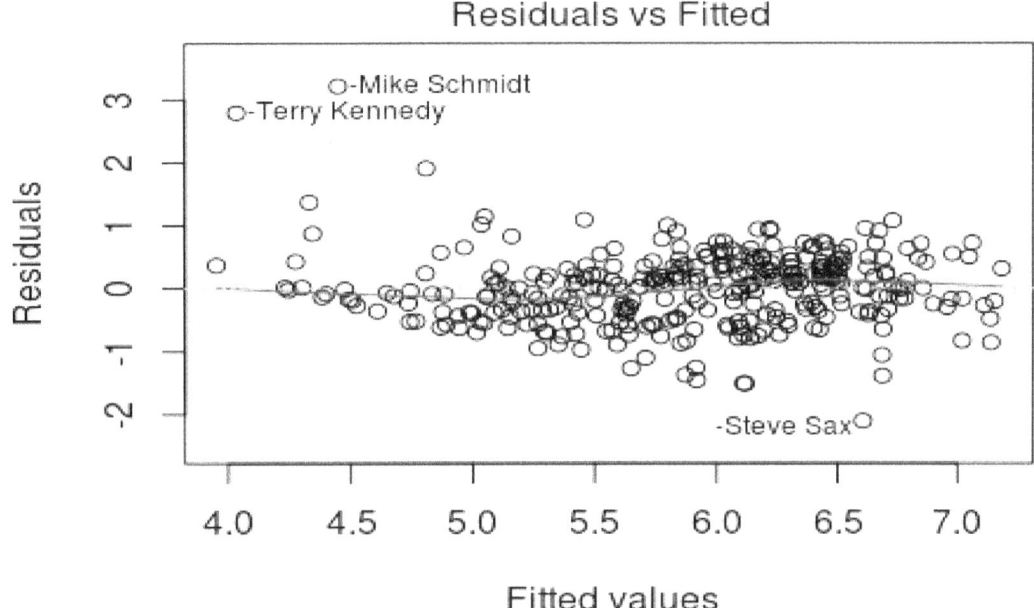

Figure 1.5: Residual vs. fitted

The residual vs. fitted plot (figure 1.5) helps us to see if there are any issues with homoscedasticity. Normally, there should be no pattern in the dispersion. If there is a pattern, it indicates that the homoscedasticity assumption is violated. It looks as though homoscedasticity is not a problem for this model.

We can confirm the visual with a simple test called the Breusch-Pagan test. The function is "bptest" and it comes from the "ltest" package.

```
bptest(Salary_model)
##
##  studentized Breusch-Pagan test
##
## data:  Salary_model
## BP = 10.776, df = 5, p-value = 0.056
```

The model passed the test but barely. This just indicates to us how deceitful the eyes can be. Figure 1.6 is the qqplot.

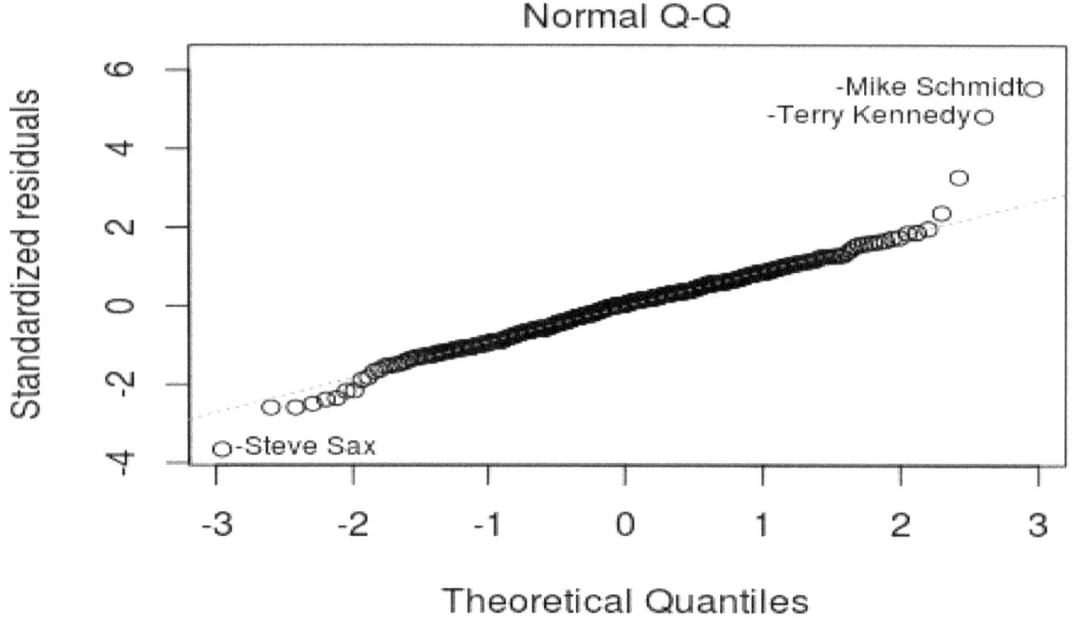

Figure 1.6: Normal QQ plot

The Q-Q plot, is a graphical tool to help us determine if a set of data plausibly came from some theoretical normal distribution. The closer the data points follow the line the better, as this indicates that the assumption of normality has not been violated. Data points that are off the line may be outliers. As you can see, our model does well with almost all of the values. However, the qqplot is a visual guess and not an absolute proof. Our next plot is figure 1.7, which is our scale-location plot.

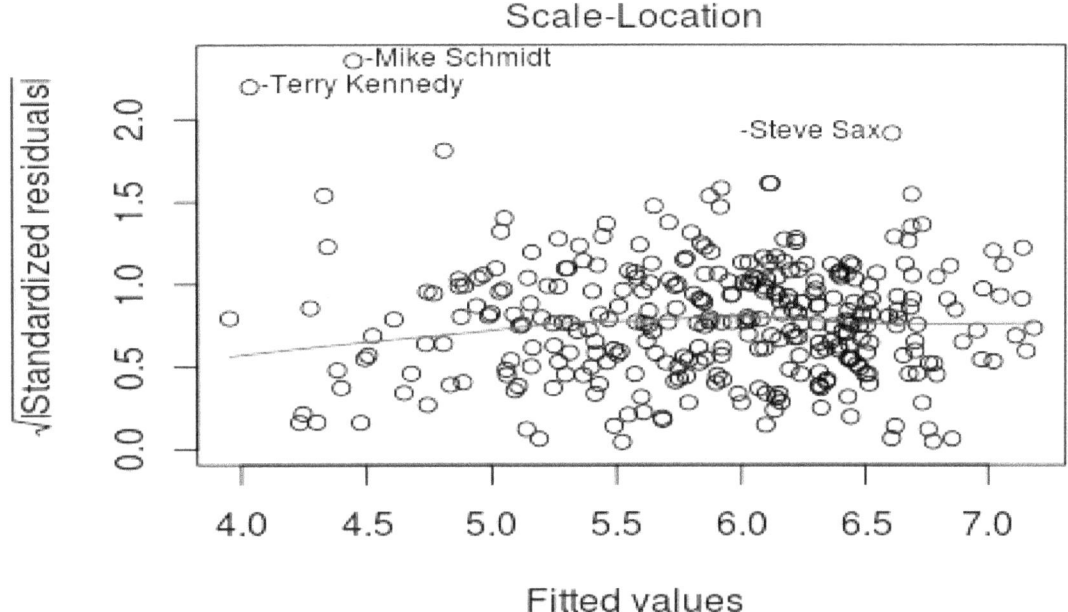

Figure 1.7: Scale location plot

The scale-location plot is similar to the residuals vs. the fitted plot (fig. 1.5). You are looking to see if there is a pattern, which is a bad sign. Again, looking at the plot, it appears our model is doing well as there is no clear pattern. Our last plot is figure 1.8, the residuals vs. leverage.

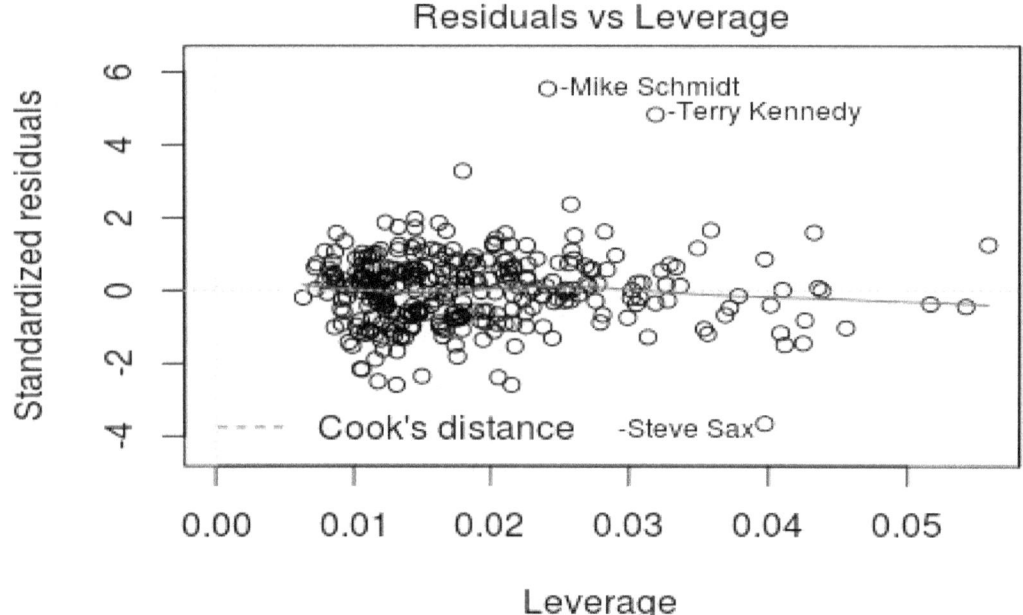

Figure 1.8: Residual vs. leverage plot

The residual vs. leverage plot can be used for locating outliers. Values greater than the absolute value of one may be outliers. Extreme values have a number next to the data point, which represents the example number from the dataset. However, in the hitters dataset the rows were given the names of the player. For example, row Mike Schmidt , Terry Kennedy, and Steve Sax all appear to be outliers.

A little bit of content knowledge will help to explain these three outliers. Mike Schmidt was usually the highest paid player in baseball for most of the 1980's even though he was on the decline physically. Terry Kennedy had an unusually high salary but his production offensively was not amazing the year the data was taken. If we had included defensive stats Terry Kennedy may not be such an outlier. Lastly, Steve Sax's salary, was low when considering his output for the year. His unusual production given his salary is why he was flagged as an outlier.

Collinearity

Our last step is to find the correlations among the variables. To do this, we need to make a correlational matrix. We need to remove variables that are not a part of our model to do this. We will use the "corrplot" function from the corrplot package to do this. Figure 1.9 is the plot.

```
completedData1<-completedData;completedData1$Chits<-
NULL;completedData1$CAtBat<-NULL;completedData1$CHmRun<-
NULL;completedData1$CRuns<-NULL;completedData1$CRBI<-
```

```
NULL;completedData1$CWalks<-NULL;completedData1$League<-
NULL;completedData1$Division<-NULL;completedData1$PutOuts<-
NULL;completedData1$Assists<-NULL; completedData1$NewLeague<-
NULL;completedData1$AtBat<-NULL;completedData1$Runs<-
NULL;completedData1$RBI<-NULL;completedData1$Errors<-NULL;
completedData1$CHits<-NULL;completedData1$Years<-NULL;
completedData1$Salary<-NULL
#remove variables that are not in the model
corrplot(cor(completedData1),method = 'number')
```

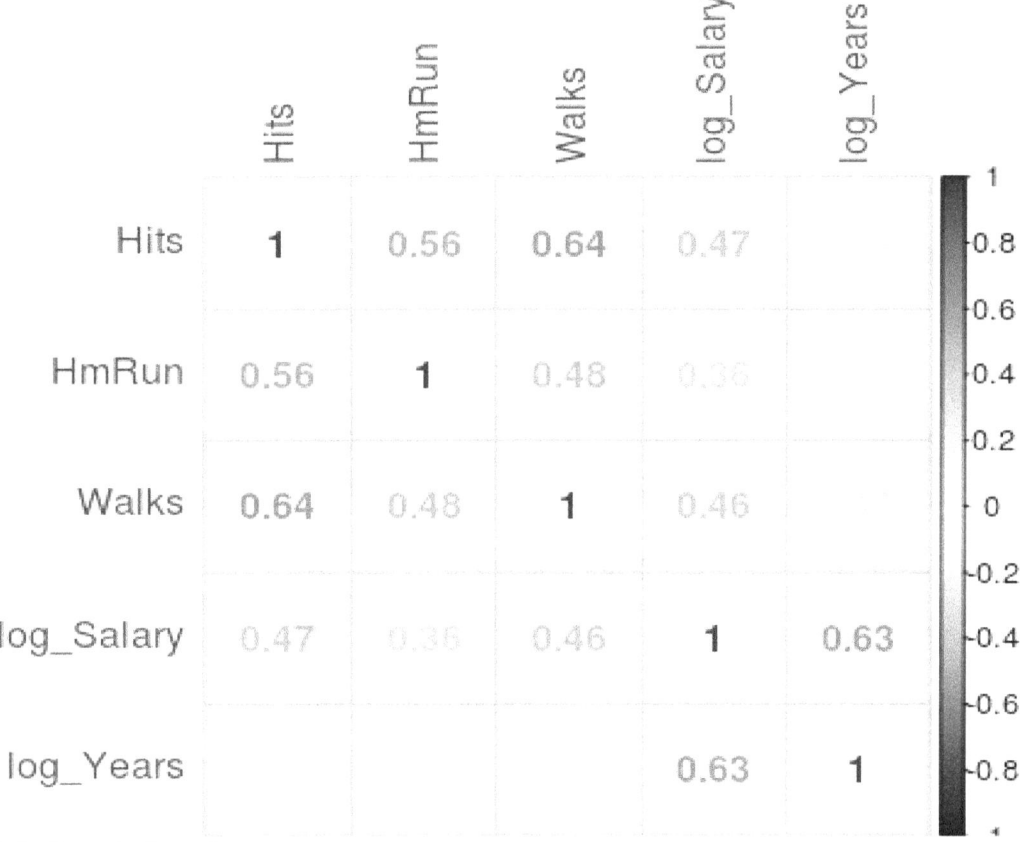

Figure 1.9: Correlation plot

There are no high correlations among our variables so collinearity is not an issue. To check for multicollinearity we need to calculate the variance inflation factor. This is done with the "vif" function from the car package.

```
vif(fit2)
##  Hits      HmRun     Walks  log_Years    League
 1.981789  1.557073  1.795250  1.039891   1.037142
```

Any values above five are considered an issue. As such, there appears to be no issues with multicollinearity.

Model Results

        With all the focus on assumptions, we never actually looked at the model outputs. Below is the code followed by an interpretation.

```
summary(Salary_Model)
##
## Call:
## lm(formula = log_Salary ~ Hits + HmRun + Walks + log_Years +
##     League, data = completedData)
##
## Residuals:
##     Min      1Q  Median      3Q     Max
## -2.1052 -0.3649  0.0171  0.3429  3.2139
##
## Coefficients:
##               Estimate Std. Error t value Pr(>|t|)
## (Intercept) 3.8790683  0.1098027  35.328  < 2e-16 ***
## Hits        0.0049427  0.0009928   4.979 1.05e-06 ***
## HmRun       0.0081890  0.0046938   1.745  0.08202 .
## Walks       0.0063070  0.0020284   3.109  0.00205 **
## log_Years   0.6390014  0.0429482  14.878  < 2e-16 ***
## League2     0.1217445  0.0668753   1.820  0.06963 .
## ---
## Signif. codes:  0 '***' 0.001 '**' 0.01 '*' 0.05 '.' 0.1 ' ' 1
##
## Residual standard error: 0.5869 on 316 degrees of freedom
## Multiple R-squared:  0.5704, Adjusted R-squared:  0.5636
## F-statistic: 83.91 on 5 and 316 DF,  p-value: < 2.2e-16
```

The model explains 57% of the variance in salary. All variables (Hits, HmRun, Walks, Years, and League) are significant at 0.1.

Conclusion

        This post provided an example dealing with missing data, checking the assumptions of a regression model, and displaying plots. In the future, we will not spend as much time checking assumptions but will move directly to model development.

Chapter Two: Best Subset Regression

In this chapter, we cover how to complete a best subset regression. Best subset regression fits a model for all possible combination of the independent variables and this is known as feature selection. For example, imagine. You have three variables.

- Income-Dependent variable
- Educational level-Independent variable
- Gender-Independent variable

With these three variables above, subset regression will create three different models as shown in the equations below

Model 1: Income ~ Educational level + Gender
Model 2: Income ~ Educational level
Model 3: Income ~ Gender

The decision for the most appropriate model is based on judgment or some statistical criteria, as we will see later. Best subset regression is an alternative to both Forward and Backward stepwise regression. Forward stepwise selection adds one variable at a time based on the lowest residual sum of squares until no additional variables lower the residual sum of squares. Backward stepwise regression starts with all variables in the model and removes variables one at a time. The concern with stepwise methods is they can produce biased regression coefficients, conflicting models, and inaccurate confidence intervals.

Best subset regression bypasses the stepwise method by creating all models possible and then allowing you to assess which variables should be including in your final model. One drawback to best subset is that a large number of variables mean a large number of potential models. With this in mind, below are the objectives of this chapter.

Chapter Objectives
- To develop an initial model using linear regression
- To develop a statistical model using best subset regression

- To select the appropriate number of variables to include in the final regression model
- To check whether our model meets the assumptions of multiple regression

Initial Model Development

We will use the "Fair" dataset from the "Ecdat" package to predict marital satisfaction based on age, Sex, the presences of children, years married, religiosity, education occupation, and number of affairs in the past year. Below is some initial code. If you do not have some of the packages in the code below be sure to install them using the "install.packages" function.

```
library(leaps);library(Ecdat);library(car);library(lmtest)
data(Fair)
```
We begin our analysis by looking at the characteristics of the variables in the dataset.
```
str(Fair)
## 'data.frame':    601 obs. of  9 variables:
##  $ sex       : Factor w/ 2 levels "female","male": 2 1 1 2 2 1 1 2 1 2 ...
##  $ age       : num  37 27 32 57 22 32 22 57 32 22 ...
##  $ ym        : num  10 4 15 15 0.75 1.5 0.75 15 15 1.5 ...
##  $ child     : Factor w/ 2 levels "no","yes": 1 1 2 2 1 1 1 2 2 1 ...
##  $ religious : int  3 4 1 5 2 2 2 2 4 4 ...
##  $ education : num  18 14 12 18 17 17 12 14 16 14 ...
##  $ occupation: int  7 6 1 6 6 5 1 4 1 4 ...
##  $ rate      : int  4 4 4 5 3 5 3 4 2 5 ...
##  $ nbaffairs : num  0 0 0 0 0 0 0 0 0 0 ...
```

You can see for yourself how many variables and the number of observations. "rate" will be the dependent variable. All other variables will be independent variables in our model. If you want a better description of each variable you can type "??Ecdat::Fair" into the console and R will tell you more about each variable.

Below is the code for our initial linear regression model. This will serve as our baseline when we do the best subset regression analysis.

```
fit<-lm(rate~.,Fair)
summary(fit)
##
## Call:
## lm(formula = rate ~ ., data = Fair)
##
## Residuals:
##     Min      1Q  Median      3Q     Max
## -3.2049 -0.6661  0.2298  0.7705  2.2292
##
## Coefficients:
##             Estimate Std. Error t value Pr(>|t|)
## (Intercept)  3.522875   0.358793   9.819  < 2e-16 ***
## sexmale     -0.062281   0.099952  -0.623  0.53346
## age         -0.009683   0.007548  -1.283  0.20005
## ym          -0.019978   0.013887  -1.439  0.15079
```

```
## childyes    -0.206976   0.116227   -1.781  0.07546 .
## religious    0.042142   0.037705    1.118  0.26416
## education    0.068874   0.021153    3.256  0.00119 **
## occupation  -0.015606   0.029602   -0.527  0.59825
## nbaffairs   -0.078812   0.013286   -5.932 5.09e-09 ***
## ---
## Signif. codes:  0 '***' 0.001 '**' 0.01 '*' 0.05 '.' 0.1 ' ' 1
##
## Residual standard error: 1.03 on 592 degrees of freedom
## Multiple R-squared:  0.1405, Adjusted R-squared:  0.1289
## F-statistic:  12.1 on 8 and 592 DF,  p-value: 4.487e-16
```

The initial results are already interesting even though the r-square is low (14%). When couples have children, they have less martial satisfaction than couples without children do when controlling for the other factors and this is the strongest regression weight. In addition, the more education a person has there is an increase in marital satisfaction. Lastly, as the number of affairs increases there is also a decrease in martial satisfaction. Keep in mind that the "rate" variable goes from 1 to 5 with one meaning a terrible marriage to five being a great one. The mean marital satisfaction was 3.52 when controlling for the other variables. This model that we just developed is the baseline model. Any model we develop after this will be consider good or bad based on its ability to improve on the numbers in this model.

Subset Regression Model
We will now create our subset models.

```
sub.fit<-regsubsets(rate~.,Fair)
best.summary<-summary(sub.fit)
```

In the code above, we created the sub models using the "regsubsets" function from the "leaps" package and saved it in the variable called "sub.fit". We then saved the summary of "sub.fit" in the variable "best.summary". We will use the "best.summary" variable several times to determine which model to use. If you call "best.summary" in the console, the output will not mean much to the average r user. However, the value of this information will be much clearer in a few pages.

Selected Variables
There are many different ways to assess a best subset regression model. We will use the following statistical methods
- Mallow' Cp
- Bayesian Information Criteria

We will make two charts for each of these criteria above. The first plot will explain how many features to include in the model. The second plot will tell you which features to include. It is important to note that for both of these methods, the lower the score for Mallow's Cp and

Bayesian Information Criteria the better the model. Below is the code for Mallow's Cp followed by figure 2.1, which indicates how many features to include in our final model.

```
plot(best.summary$cp)
plot(sub.fit,scale = "Cp")
```

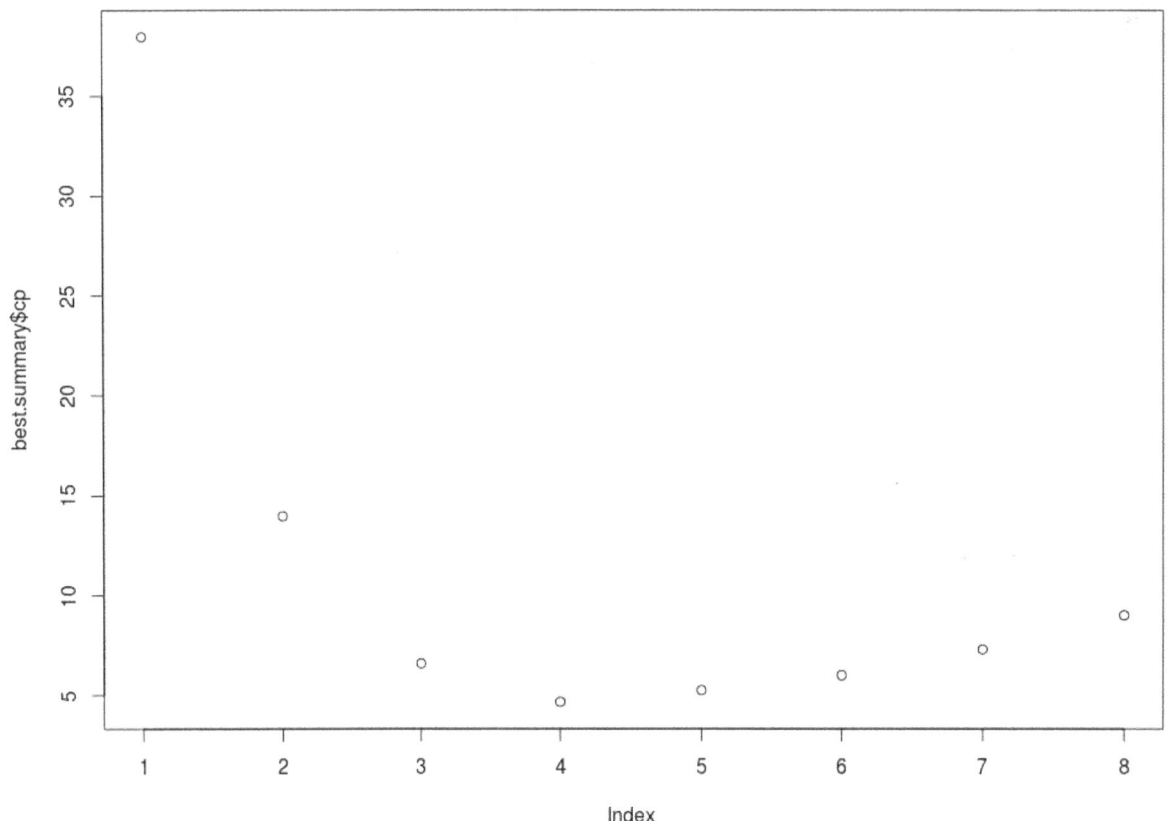

Figure 2.1: Mallow's Cp for number of features to include

Based Figure 2.1 suggests that a four-feature model is the most appropriate. However, it does not tell us which four features to include. Figure 2.2 will provide us with information on deciding which four features to include in the final model.

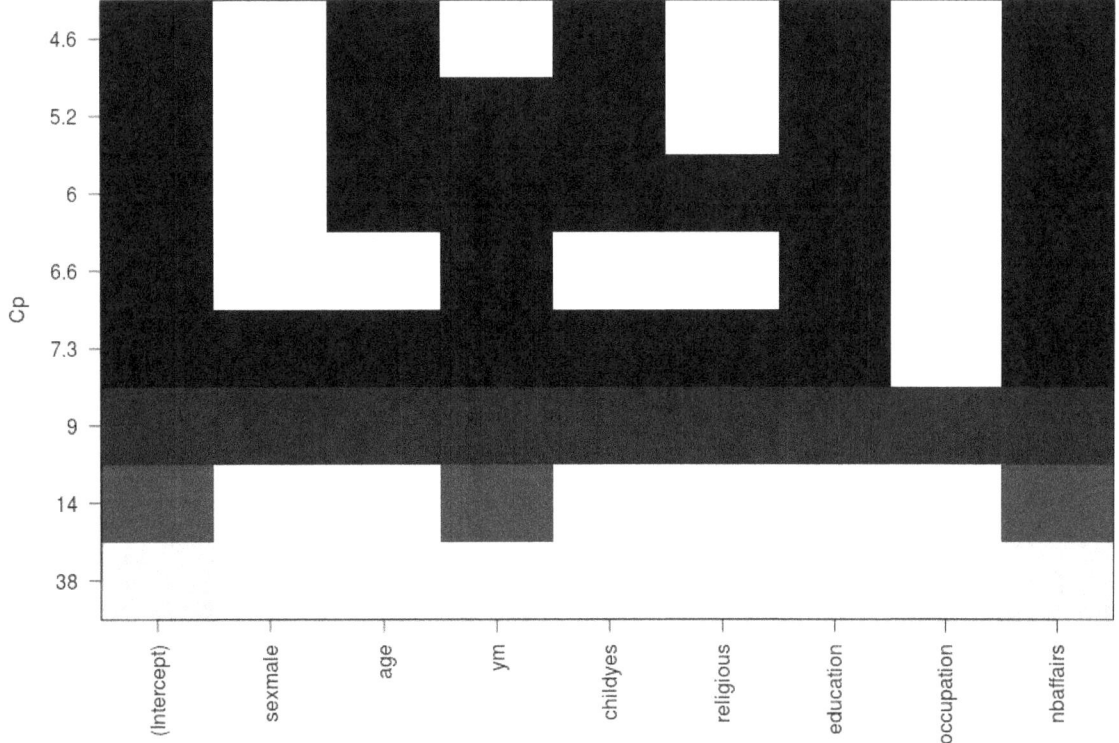

Figure 2.2 Feature Selection Mallow's Cp

Figure 2.2 reads from top to bottom rather than from bottom to top. In other words, the higher numbers are at the bottom and the lower numbers are at the top when looking at the y-axis. Knowing this, we can conclude that the most appropriate variables to include in the model are "age", "children presence", "education", and "number of affairs". Below are the results using the Bayesian Information Criterion. Figure 2.3 tells us how many features to include and figure 2.4 tells us which features.

```
plot(best.summary$bic)
plot(sub.fit,scale = "bic")
```

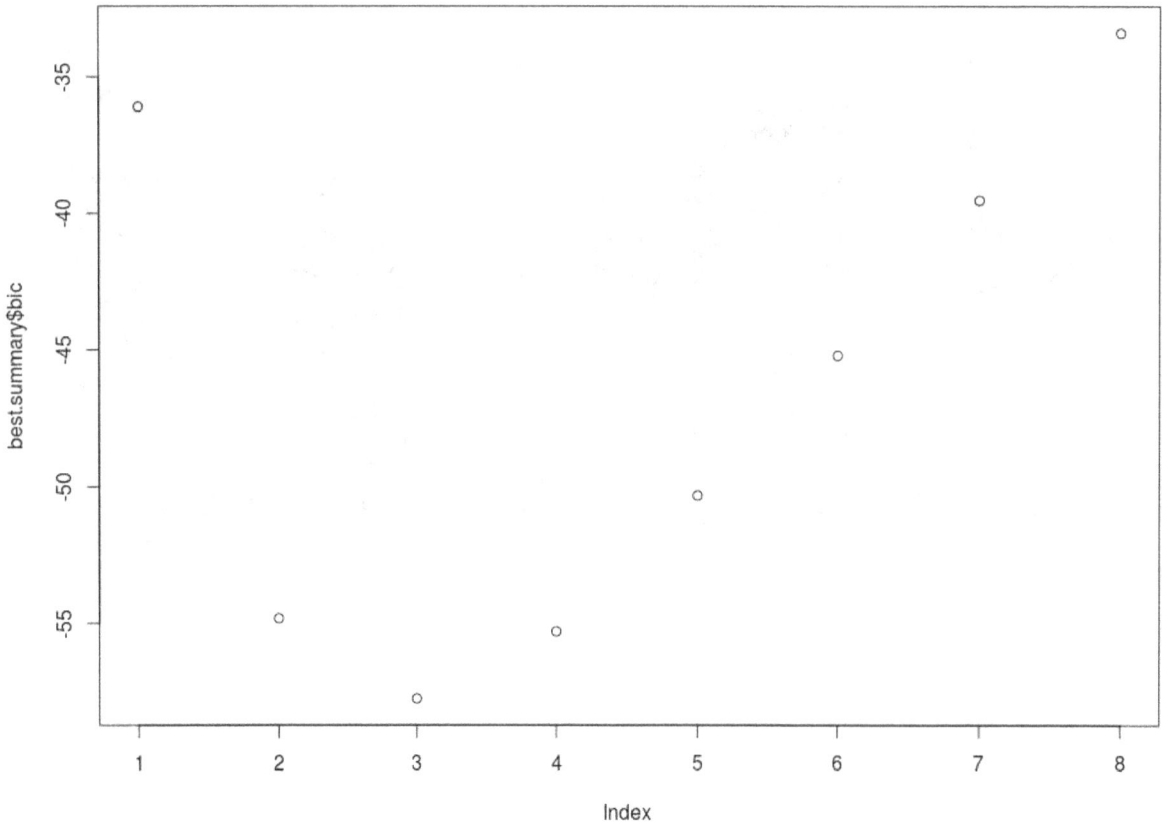

Figure 2.3: Bayesian Information Criteria for number of features to include

These results indicate that a three-feature model is appropriate. Figure 2.4 will indicate which three features to consider for the final model.

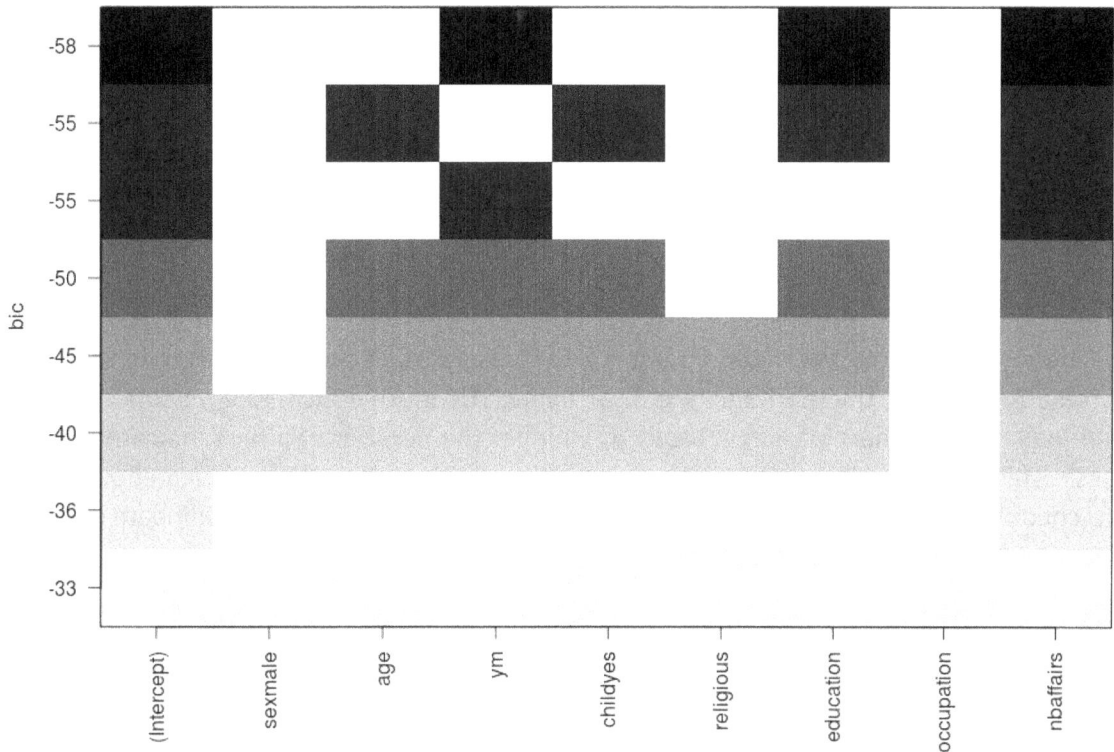

Figure 2.4 Feature Selection Bayesian Information Criteria

The variables or features to consider including are years married, education, and number of affairs. Presence of children was not considered beneficial. Since our original linear regression model indicated that presence of children was significant we will include it for now. However, whether to include this or not depends on domain knowledge and the original objectives for developing the model.

Final Model

Below is the code for the final linear regression model based on the subset regression.

```
fit2<-lm(rate~age+child+education+nbaffairs,Fair)
summary(fit2)
##
## Call:
## lm(formula = rate ~ age + child + education + nbaffairs, data = Fair)
##
## Residuals:
##     Min      1Q   Median      3Q      Max
## -3.2172 -0.7256   0.1675  0.7856   2.2713
##
```

```
## Coefficients:
##                Estimate Std. Error t value Pr(>|t|)
## (Intercept)   3.861154    0.307280  12.566  < 2e-16 ***
## age          -0.017440    0.005057  -3.449 0.000603 ***
## childyes     -0.261398    0.103155  -2.534 0.011531 *
## education     0.058637    0.017697   3.313 0.000978 ***
## nbaffairs    -0.084973    0.012830  -6.623 7.87e-11 ***
## ---
## Signif. codes:  0 '***' 0.001 '**' 0.01 '*' 0.05 '.' 0.1 ' ' 1
##
## Residual standard error: 1.029 on 596 degrees of freedom
## Multiple R-squared:  0.1352, Adjusted R-squared:  0.1294
## F-statistic: 23.29 on 4 and 596 DF,  p-value: < 2.2e-16
```

The results look ok. The older a person is the less satisfied they are with their marriage. If children are presence, the marriage is less satisfying. The more educated a person is the more satisfied they are with their marriage. Lastly, the higher the number of affairs indicates less marital satisfaction. This all seems to make practical sense. However, before we get excited we need to check for collinearity and homoscedasticity. First, we will check for collinearity use the "vif" function from the "car" package. Below is the code

```
vif(fit2)
##       age     child education  nbaffairs
##  1.249430  1.228733  1.023722   1.014338
```

According to the vif results, there are no issues with collinearity, as the values should exceed 5 or 10.

We will now check for homoscedasticity. To do this we will use the "plot" function and insert are "fit2" model as an argument. This will generate four plots that each provides insights into whether we have met the assumptions of linear regression. Figure 2.5 is the residual vs. fitted plot and it is the first one shown below.

```
plot(fit2)
```

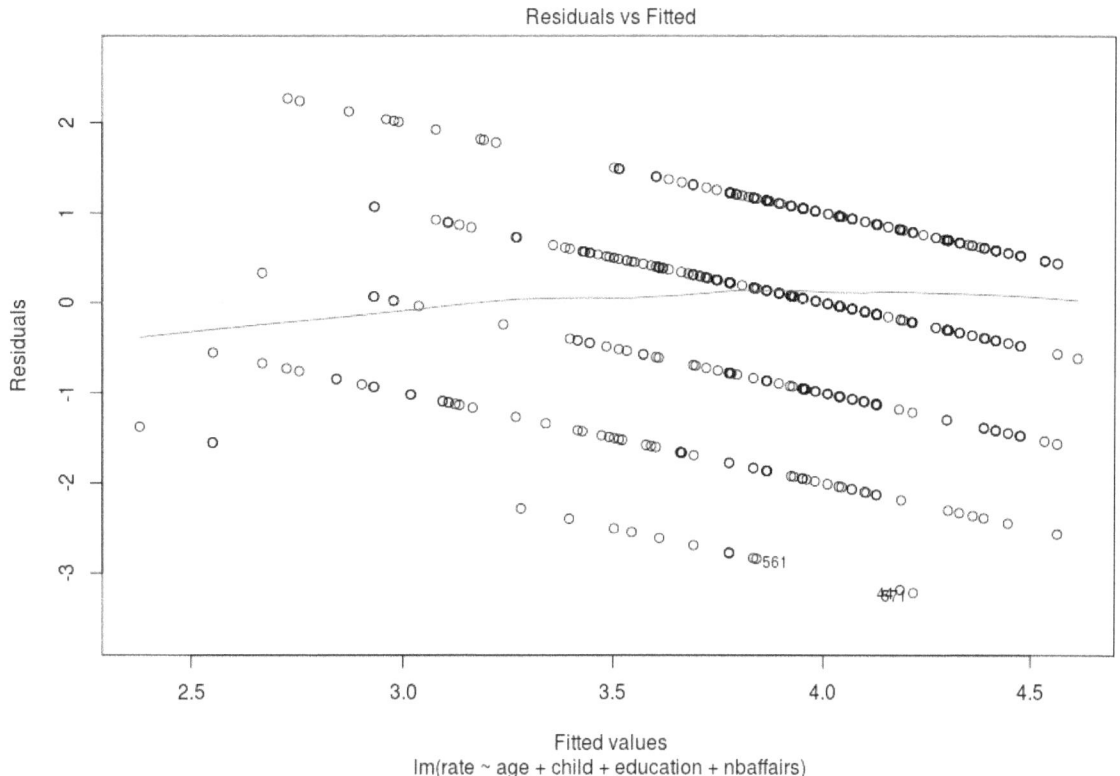

Figure 2.5: Residual Vs Fitted Plot

Figure 2.5, the residuals vs. the fitted plot does not look good. Normally, there should be no pattern in the dispersion. If there is a pattern, it indicates that homoscedasticity assumption is violated. Figure 2.6 is the qqplot

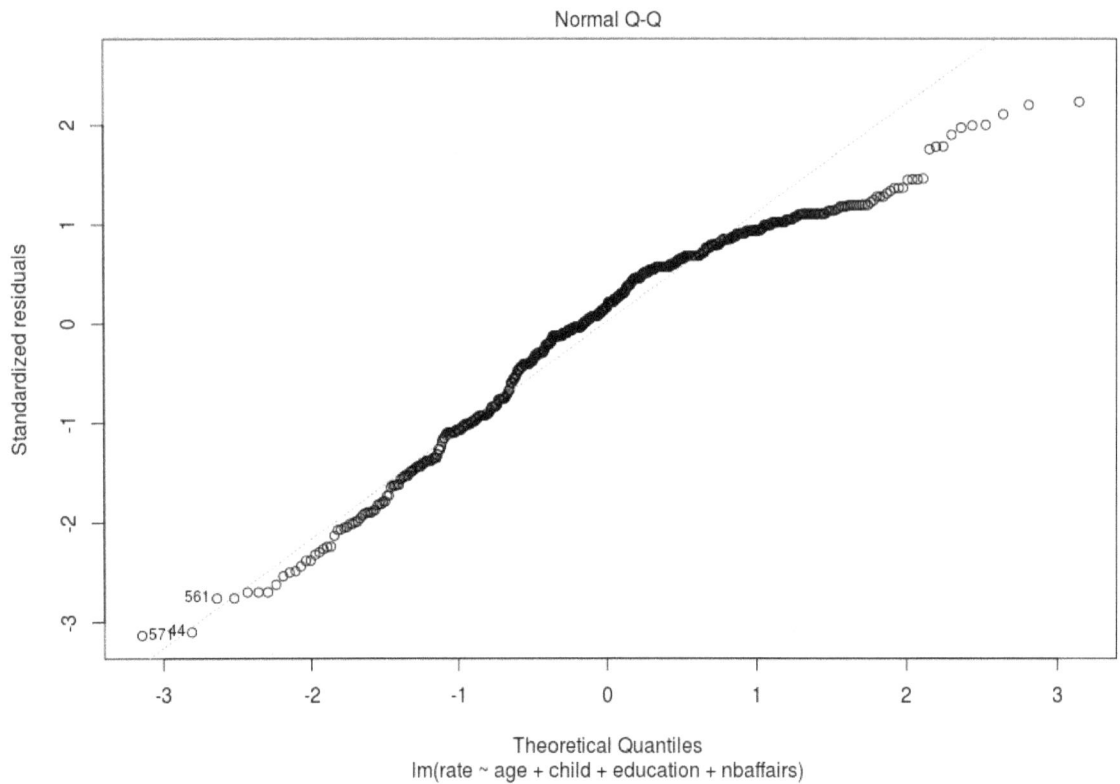

Figure 2.6: QQ plot

The Q-Q plot, is a graphical tool to help us determine if a set of data plausibly came from some theoretical normal distribution. The closer the data points follow the line the better. Data points that are off the line may be outliers. As you can see, our model does well with lower values but struggles to estimate higher values. The departure may be strong enough to indicate our model is not doing a well. However, the qqplot is a visual guess and not absolute proof. Our next plot is figure 2.7, which is our scale-location plot.

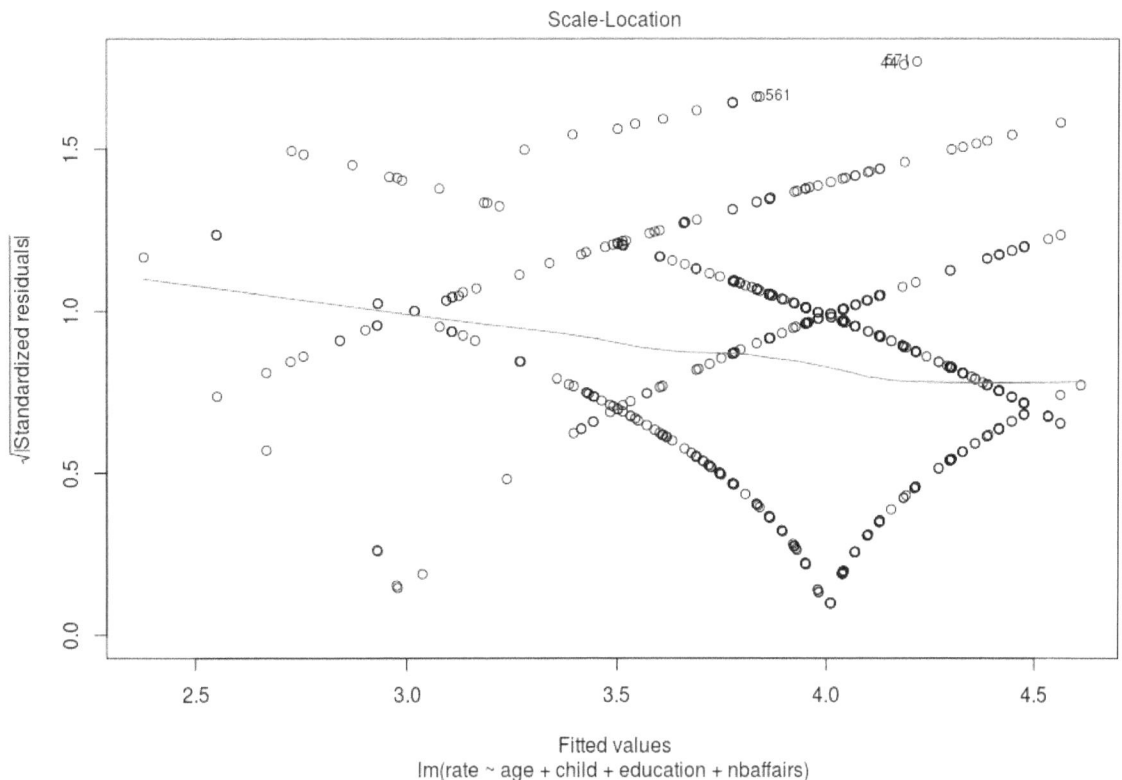

Figure 2.7: Scale-Location

The scale-location plot is similar to the residuals vs. the fitted plot (fig. 2.5). You are looking to see if there is a pattern, which is a bad sign. Again, looking at the plot, it appears our model is in trouble and that there is a clear pattern in the data. Our last plot is figure 2.8, the residuals vs. leverage.

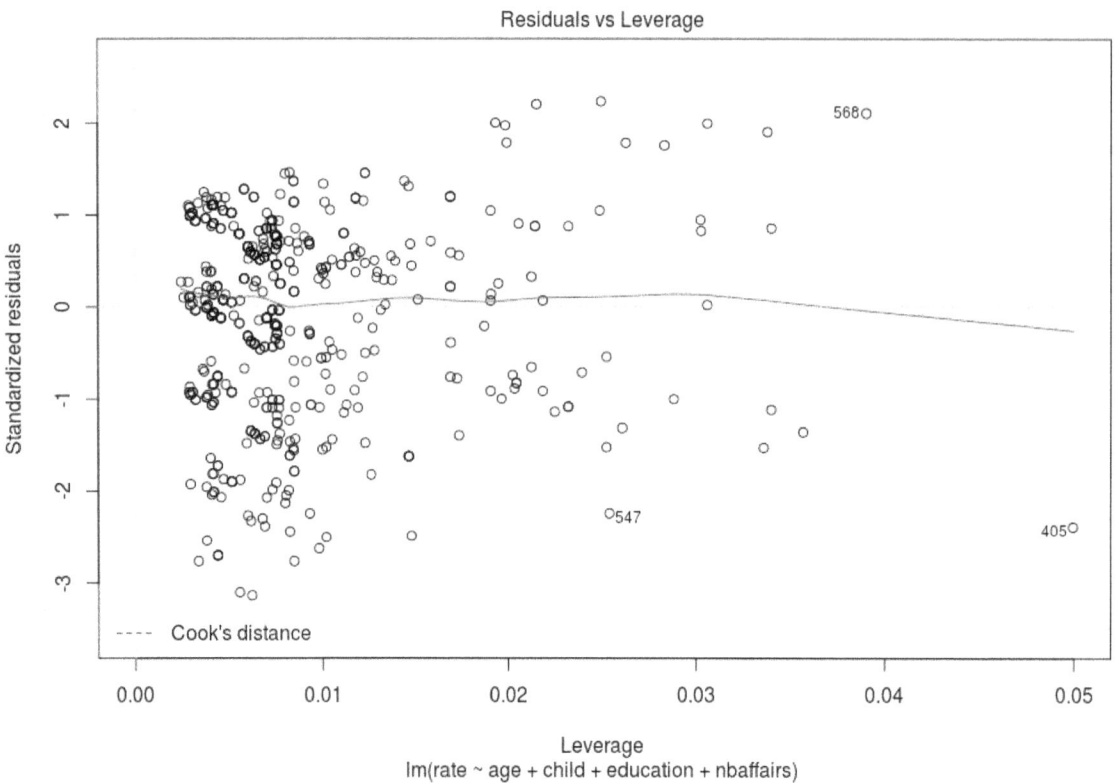

Figure 2.8: Residual vs. leverage

The residual vs. leverage plot can be used for locating outliers. Values greater than the absolute value of 1 may be outliers. Extreme values have a number next to the data point, which represents the example number from the dataset. For example, row 547 appears to be an outlier as well as row 405 and 568.

Our model seems to be poor based on visuals. However, we need to confirm uses statistics and not just our eyes to be sure. This will be done by using the Breusch-Pagan test from the "lmtest" package. Below is the code

```
bptest(fit2)
##
##  studentized Breusch-Pagan test
##
## data:  fit2
## BP = 16.238, df = 4, p-value = 0.002716
```

There you have it. Our model violates the assumption of homoscedasticity. However, this model was developed for demonstration purposes to provide an example of subset regression.

Conclusion

Best subset regression allows you to use statistical analysis to determine which features to include in your final model. This approach has the advantage of reducing bias and inaccurate results. However, best subset strength is in developing models that do not have a large number of features. If there are too many features the "regsubset" function will not conduct an analysis. When the features are many it is necessary to use regularized regression techniques.

Chapter Three: Ridge Regression

Traditional linear regression is a tried and true model for making predictions for decades. Best subset regression was a step forward in trying to select the best features to include in a regression model. However, with the growth of Big Data and datasets with 100's of variables problems have begun to arise in terms of selecting features as well as running models with millions of examples. With these challenges present, using stepwise or best subset method with regression could take hours if not days to converge in even some of the best computers.

To deal with this problem, regularized regression has been developed to help to determine which features or variables to keep when developing models from large datasets with a huge number of variables. However, we need to make sure we understand what regularization means and its relation to regression. In this chapter, we will do the following...

Chapter Objectives
- Define regularization
- Explain the mechanics of ridge regression
- Prepare a data set for ridge regression analysis
- Develop a model using ridge regression
- Test a model developed using ridge regression

Regularization

Regularization involves the use of a shrinkage penalty in order to reduce the residual sum of squares (RSS). This is done through selecting a value for a tuning parameter called "lambda". Tuning parameters are used in machine learning algorithms to control the behavior of the models that are developed.

The lambda is multiplied by the normalized coefficients of the model and added to the RSS. Below is an equation of what was just said

$$RSS + \lambda(\text{normalized coefficients})$$

The benefits of regularization are at least three-fold. First, regularization is highly computationally efficient. Instead of fitting k-1 models when k is the number of variables available (for example, 50 variables would lead 49 models!), with regularization only one model is developed for each value of lambda you specify.

Second, regularization helps to deal with the bias-variance headache of model development. When small changes are made to data, such as switching from the training to testing data, there can be wild changes in the estimates. Regularization can often smooth this problem out substantially.

Finally, regularization can help to reduce or eliminate any multicollinearity in a model. As such, the benefits of using regularization make it clear that this should be considering when working with larger data sets and or a large number of features.

There are at least three types of regularization, ridge, lasso, and elastic net. In this chapter, we will take a closer look at ridge regression and will deal with both lasso and elastic net in future chapters. Below is an explanation of ridge regression.

Ridge Regression

Ridge regression involves the normalization of the squared weights or as shown in the equation below

$$\text{Residual sum of squares} + \lambda(\text{normalized coefficients}^2)$$

This is also referred to as the L2-norm. As lambda (λ) increases in value, the coefficients in the model are shrunk towards 0 but never reach 0 (shrinkage penalty). This is how the error is shrunk. The higher the lambda the lower the value of the coefficients as they are reduce more and more thus reducing the RSS.

The benefit is that predictive accuracy is often increased. However, interpreting and communicating your results can become difficult because no variables are removed from the model. Instead, the variables are reduced near to zero. This can be especially tough if you have dozens of variables remaining in your model to try to explain.

Data Preparation

We will now conduct an analysis using ridge regression. We will use the "SAheart" dataset form the "ElemStatLearn" package. We want to predict systolic blood pressure (sbp) using all of the other variables available as predictors. Below is some initial code that we need in order to begin.

```
library(ElemStatLearn);library(car);library(corrplot)
library(leaps);library(glmnet);library(caret)
data(SAheart)
str(SAheart)
## 'data.frame':    462 obs. of  10 variables:
##  $ sbp      : int  160 144 118 170 134 132 142 114 114 132 ...
##  $ tobacco  : num  12 0.01 0.08 7.5 13.6 6.2 4.05 4.08 0 0 ...
##  $ ldl      : num  5.73 4.41 3.48 6.41 3.5 6.47 3.38 4.59 3.83 5.8 ...
##  $ adiposity: num  23.1 28.6 32.3 38 27.8 ...
```

```
##  $ famhist  : Factor w/ 2 levels "Absent","Present": 2 1 2 2 2 2 1 2 2 2
...
##  $ typea    : int   49 55 52 51 60 62 59 62 49 69 ...
##  $ obesity  : num   25.3 28.9 29.1 32 26 ...
##  $ alcohol  : num   97.2 2.06 3.81 24.26 57.34 ...
##  $ age      : int   52 63 46 58 49 45 38 58 29 53 ...
##  $ chd      : int   1 1 0 1 1 0 0 1 0 1 ...
```

A look at the object using the "str" function indicates that one variable "famhist" is a factor variable. The "glmnet" function that does the ridge regression analysis cannot handle factors so we need to converts this to a dummy variable. However, there are two things we need to do before this. First, we need to check the correlations to make sure there are no major issues with multicollinearity. Second, we need to create our training and testing data sets. Below is the code for the correlation plot followed by figure 4.1 the correlation plot

```
p.cor<-cor(SAheart[,-5])
corrplot.mixed(p.cor)
```

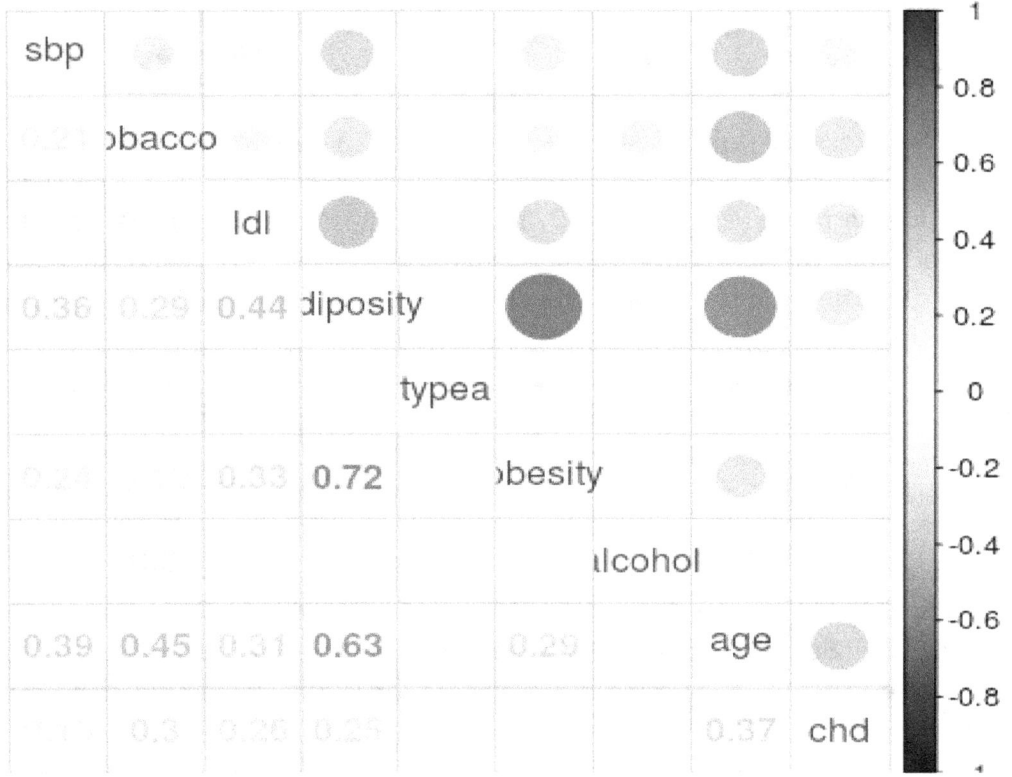

Figure 3.1: Correlation plot

First, we created a variable called "p.cor" the -5 in brackets means we removed the 5th column from the "SAheart" data set, which is the factor variable "Famhist". The correlation plot indicates that there is one strong relationship between adiposity and obesity. However, one common cut-off for collinearity is 0.8 and this value is 0.72, which is not a problem.

We will now create our training and testing sets and convert "famhist" to a dummy variable.

```
#create train and test datasets
ind<-sample(2,nrow(SAheart),replace=T,prob = c(0.7,0.3))
train<-SAheart[ind==1,]
test<-SAheart[ind==2,]
train$famhist<-model.matrix( ~ famhist - 1, data=train )
#convert to dummy variable
test$famhist<-model.matrix( ~ famhist - 1, data=test )
```

We are still not done preparing our data yet. "glmnet" cannot use data frames, instead, it can only use matrices. Therefore, we now need to convert our data frames to matrices. We have to create two matrices, one with all of the predictor variables and a second with the outcome variable of blood pressure. Below is the code

```
predictor_variables<-as.matrix(train[,2:10])
blood_pressure<-as.matrix(train$sbp)
```

Model Development

We are now ready to create our model. We use the "glmnet" function and insert our two matrices. The 'family' argument is set to Gaussian because "blood pressure" is a continuous variable. Alpha is set to 0 as this indicates ridge regression (more on this later). Below is the code

```
ridge<-glmnet(predictor_variables,blood_pressure,family = 'gaussian',alpha = 0)
```

Now we need to look at the results using the "print" function. This function prints a lot of information as explained below.
- Df = number of variables including in the model (this is always the same number in a ridge model)
- %Dev = Percent of deviance explained. The higher the better
- Lambda = The lambda used to attain the %Dev

When you use the "print" function for a ridge model, it will print up to 100 different models. Fewer models are possible if the percent of deviance stops improving. 100 is the default stopping point. In the code below, we have the "print" function. However, I have only printed the first 5 and last 5 models in order to save space.

```
print(ridge)
##
## Call:  glmnet(x = predictor_variables, y = blood_pressure, family =
"gaussian",        alpha = 0)
##
##         Df      %Dev      Lambda
##   [1,] 10 7.413e-37 7822.0000
##   [2,] 10 2.047e-03 7127.0000
##   [3,] 10 2.244e-03 6494.0000
```

```
##    [4,] 10 2.460e-03 5917.0000
##    [5,] 10 2.696e-03 5391.0000
.................................................................. .
##   [95,] 10 1.690e-01    1.2290
##   [96,] 10 1.691e-01    1.1190
##   [97,] 10 1.692e-01    1.0200
##   [98,] 10 1.693e-01    0.9293
##   [99,] 10 1.693e-01    0.8468
## [100,] 10 1.694e-01    0.7716
```

The results from the "print" function are useful in setting the lambda for the model when we use the test dataset. Based on the results we can set the lambda at near 0.83 because this explains the highest amount of deviance at almost .20. The results will vary slightly each time unless you set the seed use the "set.seed" function

Keep in mind that by setting the lambda you are determining the value of the coefficients in your model, while keeping all of the potential variables in it. In best subset regression, R helps you to select the features but does not determine the values of the coefficient. However, with ridge regression R keeps all the features but sets the coefficients through determining what the best lambda value is statistically.

Figure 3.2 below shows us the lambda on the x-axis and the coefficients of the predictor variables on the y-axis. The numbers inside the plot refer to the actual coefficient of a particular variable. Each number corresponds to a variable going from left to right in the dataframe/matrix using the "View" function. For example, the 1 in the plot refers to "tobacco" 2 refers to "ldl" etc. The number across the top of the graph is the number of variables used in the model. For ridge regression, this number stays the same because ridge regression never removes features.

```
plot(ridge,xvar="lambda",label=T)
```

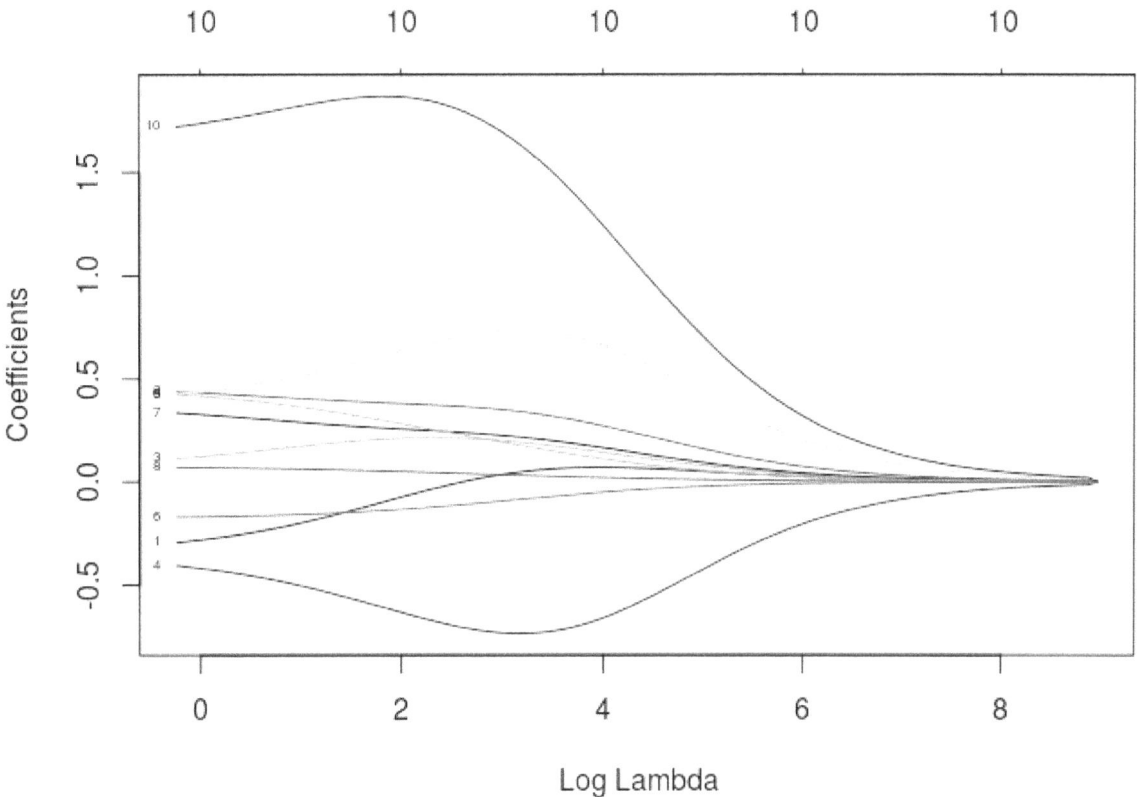

Figure 3.2 Lambda and Coefficients of the variables

As you can see, as lambda increase the coefficient decrease in value. This is how ridge regression works yet no coefficient ever goes to absolute 0.

You can also look at the coefficient values at a specific lambda value. The values are unstandardized but they provide a useful insight when determining final model selection. In the code below, the lambda is set to .83 and we use the "coef" function to do this

```
ridge.coef<-coef(ridge,s=.83,exact = T)
ridge.coef
## 11 x 1 sparse Matrix of class "dgCMatrix"
##                                1
## (Intercept)           107.09456597
## tobacco                -0.15994444
## ldl                    -0.28081041
## adiposity               0.19362474
## famhist.famhistAbsent   0.85234170
## famhist.famhistPresent -0.81760601
## typea                  -0.01272944
## obesity                 0.27937218
## alcohol                 0.03634111
```

```
## age                    0.45765671
## chd                    1.14615711
```

The information in the printout above tells us the same information we would get in a regular regression analysis except taking into account the shrinkage penalty that the lambda provides. Some of the results are strange. For example, if tobacco goes up 1 unit then blood pressure goes done .15 units. If family history does not have high blood pressure, blood pressure still goes up 0.85 units. The interpretation of the model is not the point of the chapter rather the use and application of ridge regression is.

Figure 3.3 shows us the deviance explained on the x-axis and the coefficients of the predictor variables on the y-axis. Below is the code.

```
plot(ridge,xvar='dev',label=T)
```

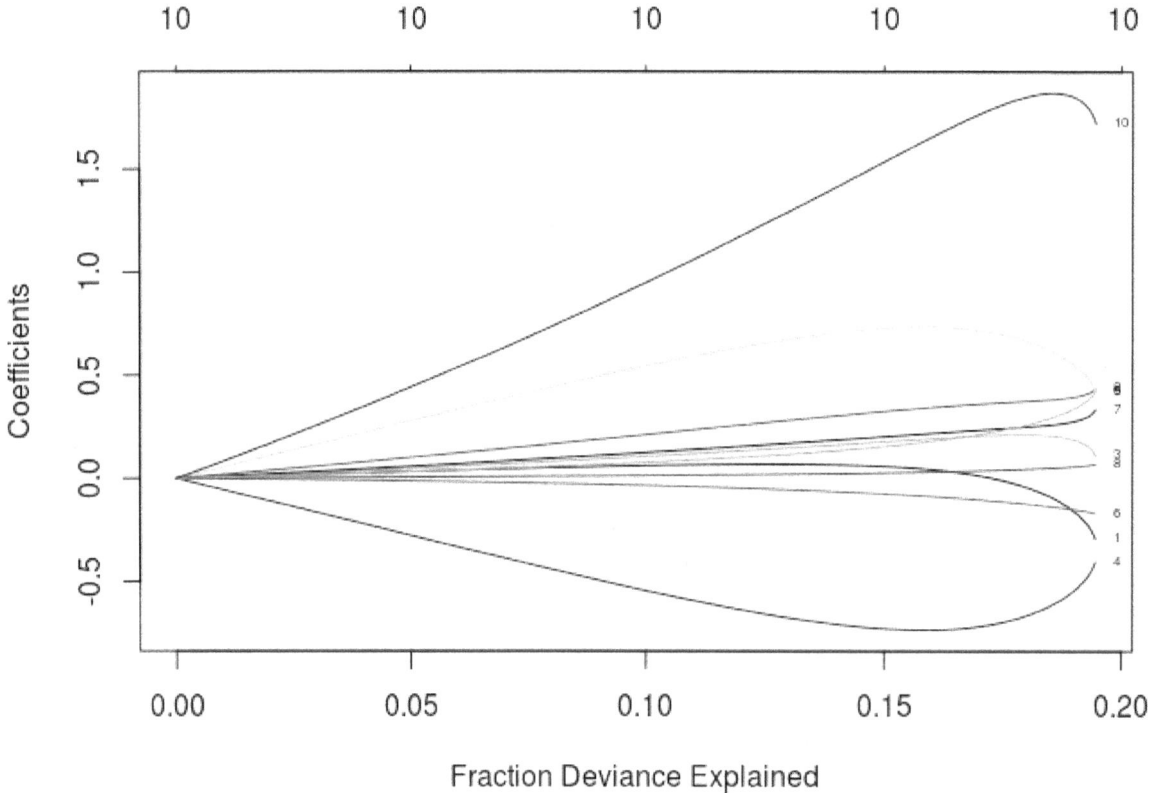

Figure 3.3: Deviance Explained

Figure 3.2 and 3.3 are completely opposite to each other. Increasing lambda causes a decrease in the coefficients while increasing the fraction of deviance explained leads to an increase in the coefficients. You can also see this when we used the "print"" function earlier to look at the different models that were developed. As lambda became smaller, there was an increase in the deviance explained.

Model Testing

We now can begin testing our model on the test data set. We need to convert the test dataset to a matrix and then we will use the "predict" function while setting our lambda to .83. Lastly, we will plot the results as shown in figure 3.4. Below is the code. The "s" argument is for setting the lambda the rest of the code should be familiar.

```
test.matrix<-as.matrix(test[,2:10])
ridge.y<-predict(ridge,newx = test.matrix,type = 'response',s=.83)
plot(ridge.y,test$sbp)
```

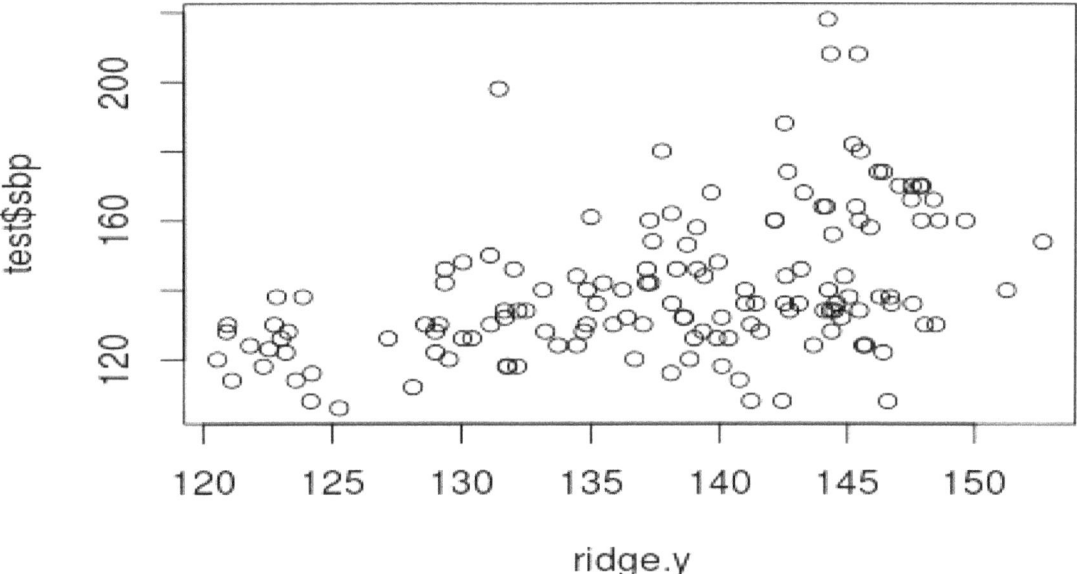

Figure 3.4: Model comparison to test set

Visually, the results do not look that impressive. However, a model needs to be compared to another model to assess its quality.

The last thing we need to do is calculate the mean squared error (MSE). By its self, this number is useless. However, it provides a benchmark for comparing the current model with any other models you may develop using other methods. Below is the code

```
ridge.resid<-ridge.y-test$sbp
mean(ridge.resid^2)
## [1] 364.8326
```

Knowing this number, we can develop other models using other methods of analysis to try and reduce it as much as possible.

Conclusion

This chapter provided an introduction to ridge regression. Ridge regression is part of a family of regularized algorithms that apply a shrinkage penalty to coefficient. These algorithms were developed in response to the feature selecting as well as computational limitations of stepwise and best subset regression.

The key aspect of ridge regression is selecting the lambda. The lambda is what affects your coefficients and it is the coefficients that determine the accuracy of your final model. Understanding how this is done is the main difference of ridge regression from traditional regression.

Chapter Four: Lasso Regression

In this post, we will conduct an analysis using the lasso regression. Remember lasso regression will actually eliminate variables by reducing them to zero through how the shrinkage penalty can be applied.

The second type of regularized regression we will look at is lasso regression. Lasso is short for "Least Absolute Shrinkage and Selection Operator". This approach uses the L1-norm, which is the sum of the absolute value of the coefficients or as shown in the equation below.

$$RSS + \lambda(\Sigma|normalized\ coefficients|)$$

This shrinkage penalty will reduce a coefficient to zero, which is another way of saying that variables were removed from the model. Remember, ridge regression never removes any variables. For lasso regression, one potential problem with removing variables is that highly correlated variables that need to be in your model may be removed when Lasso shrinks coefficients. This problem with unnecessary removal of variables is one reason why ridge regression is still used. Below are the objectives of this chapter.

Chapter Objectives
- Briefly explore and setup a dataset for analysis
- Develop a regression model using lasso
- Assess the models performance

Data Preparation
We will use the dataset "nlschools" from the "MASS" packages to conduct our analysis. We want to see if we can predict language test scores "lang" with the other available variables. Below is some initial code to begin the analysis. Keep in mind that the majority of the code in this chapter is the same as chapter 4 but we are using a different dataset.

```
library(MASS);library(corrplot);library(glmnet)
data("nlschools")
str(nlschools)
```

```
## 'data.frame':    2287 obs. of  6 variables:
## $ lang : int  46 45 33 46 20 30 30 57 36 36 ...
## $ IQ   : num  15 14.5 9.5 11 8 9.5 9.5 13 9.5 11 ...
## $ class: Factor w/ 133 levels "180","280","1082",..: 1 1 1 1 1 1 1 1 1 1
...
## $ GS   : int  29 29 29 29 29 29 29 29 29 29 ...
## $ SES  : int  23 10 15 23 10 10 23 10 13 15 ...
## $ COMB : Factor w/ 2 levels "0","1": 1 1 1 1 1 1 1 1 1 1 ...
```

We need to remove the "class" variable as it is used as an identifier and provides no useful data. After this, we can check the correlations among the variables by creating a correlation plot. Below is the code for this. Figure 4.1 is the correlation plot.

```
nlschools$class<-NULL
p.cor<-cor(nlschools[,-5]) # -5 takes removes the class variable
corrplot.mixed(p.cor)
```

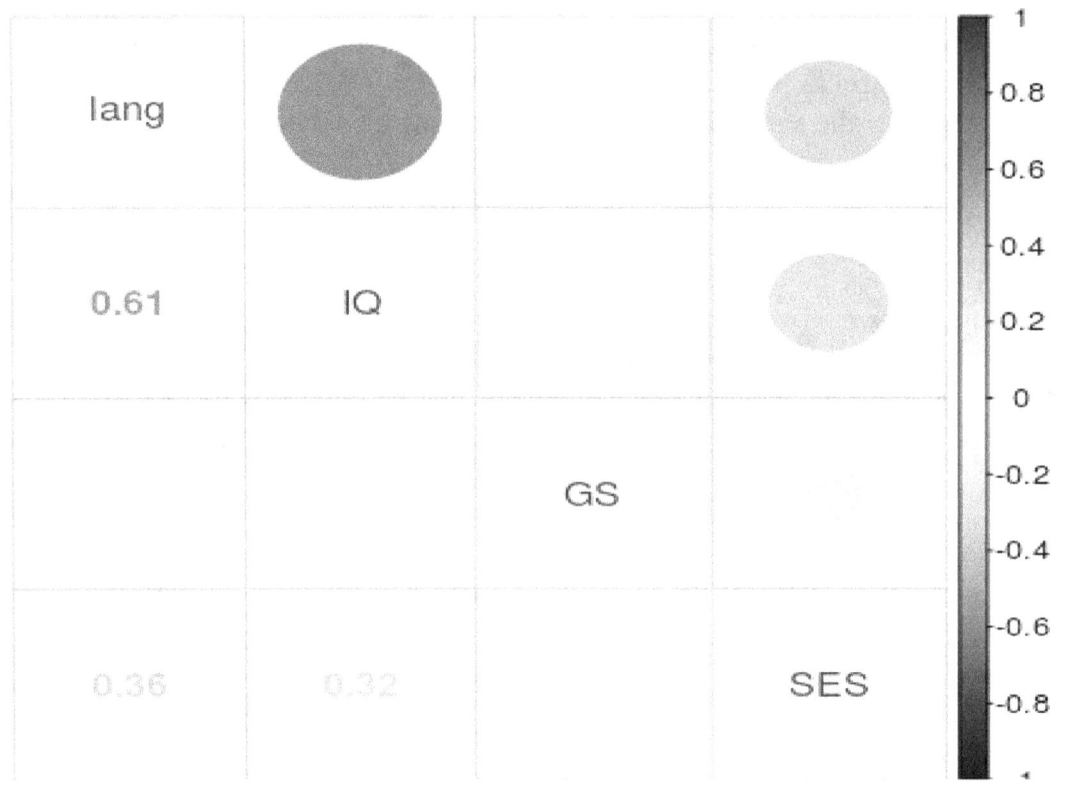

Figure 4.1: Correlation plot

There appears to be no problems with collinearity. We will now setup are training and testing sets.

```
ind<-sample(2,nrow(nlschools),replace=T,prob = c(0.7,0.3))
train<-nlschools[ind==1,]
test<-nlschools[ind==2,]
```

Remember that the 'glmnet' function does not like factor variables. Therefore, we need to convert our "COMB" variable to a dummy variable. In addition, "glmnet" function also does not like dataframes so we need to make two matrices. The predictor variables will be copied into a matrix called "predictor_variables" and the dependent variable "lang" will be placed in a matrix called "language_score." Below is the code

```
train$COMB<-model.matrix( ~ COMB - 1, data=train ) #convert to dummy variable
test$COMB<-model.matrix( ~ COMB - 1, data=test )
predictor_variables<-as.matrix(train[,2:4])
language_score<-as.matrix(train$lang)
```

Model Development

We can now run our model. We place both matrices inside the "glmnet" function. The family is set to "gaussian" because our outcome variable is continuous. The "alpha" is set to 1 as this indicates that we are using lasso regression. This point about the alpha is important. The main difference in the code of ridge and lasso regression in R is that the alpha is zero for ridge regression and one for lasso. The code follows

```
lasso<-glmnet(predictor_variables,language_score,family="gaussian",alpha=1)
```

Now we need to look at the results using the "print" function. As you may know from chapter 4, the "print" function prints a lot of information. Below is a repeat from chapter 4 of the information the "print" function supplies
- Df = number of variables including in the model (this is always the same number in a ridge model)
- %Dev = Percent of deviance explained. The higher the better
- Lambda = The lambda used to attain the %Dev

Remember, when you use the "print" function for a lasso model it will print up to 100 different models. Fewer models are possible if the percent of deviance stops improving. 100 is the default stopping point. In the code below, we will use the "print" function but I only printed the first 5 and last 5 models in order to reduce the size of the printout.

```
print(lasso)
##
## Call:  glmnet(x = predictor_variables, y = language_score, family =
"gaussian", alpha = 1)
##
##        Df    %Dev  Lambda
## [1,]  0 0.00000 5.58700
## [2,]  1 0.06547 5.09100
## [3,]  1 0.11980 4.63900
## [4,]  1 0.16490 4.22700
## [5,]  1 0.20240 3.85100
.........................................................
## [55,]  3 0.41070 0.03676
## [56,]  3 0.41080 0.03350
```

```
## [57,]  3 0.41080 0.03052
## [58,]  3 0.41080 0.02781
## [59,]  3 0.41080 0.02534
## [60,]  3 0.41080 0.02309
```

The results from the "print" function will allow us to set the lambda for the "test" dataset. Based on the results we can set the lambda at 0.02 because this explains the highest amount of deviance at .41. Remember your results may be slightly different.

Figure 4.2 below shows us lambda on the x-axis and the coefficients of the predictor variables on the y-axis. The numbers next to the coefficient lines refers to the actual coefficient of a particular variable as it changes from using different lambda values. Each number corresponds to a variable going from left to right in a dataframe/matrix using the "View" function. For example, 1 in the plot refers to "IQ" 2 refers to "GS" etc. Across the top of the graph is the number of variables in the model. Unlike ridge, in which all models the number of variables is the same, in lasso, variables are removed from the model by the shrinkage penalty. Below is the actual code and plot.

```
plot(lasso,xvar="lambda",label=T)
```

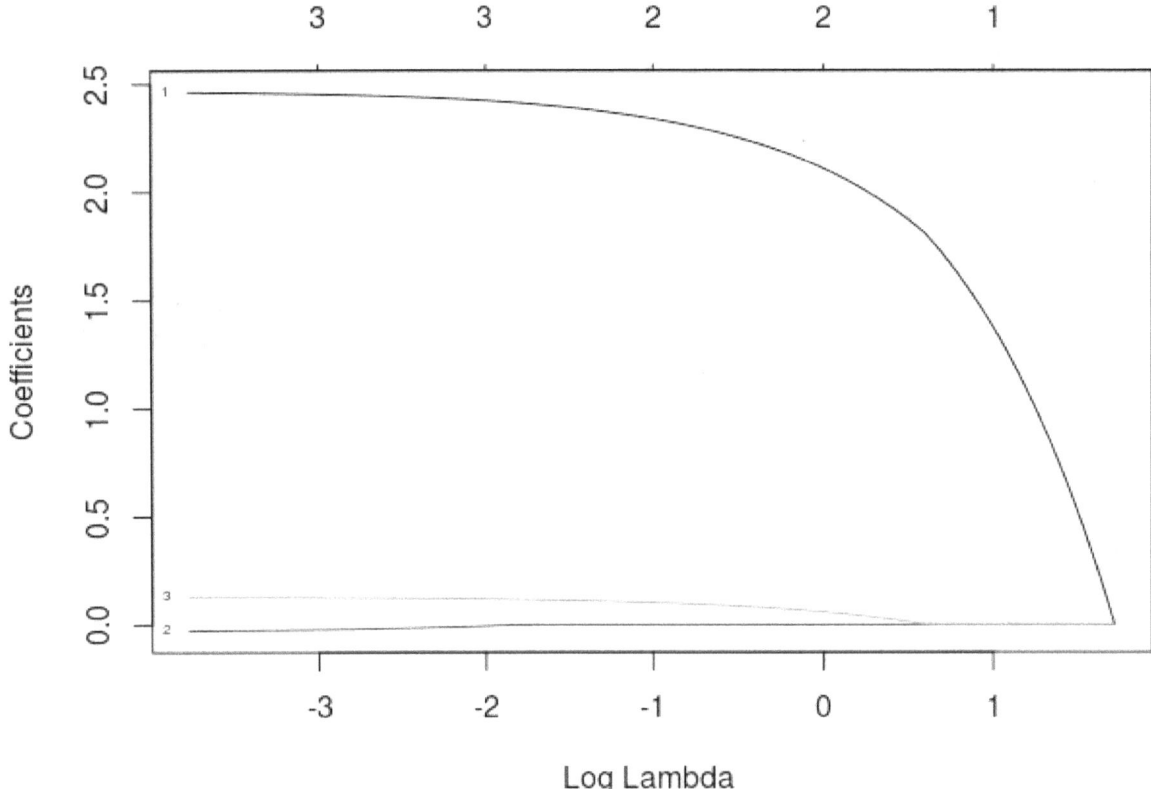

Figure 4.2: Lambda and coefficent plot

As you can see, as lambda increases the coefficients decrease in value. This is how regularized regression works. However, unlike ridge regression, which never reduces a coefficient to zero, lasso regression does reduce a coefficient to zero. For example, coefficient 3 (SES variable) and Coefficient 2 (GS variable) are reduced to zero when lambda is near 1.

You can also look at the coefficient values at a specific lambda values. The values are unstandardized and are used to determine the final model selection. In the code below, the lambda is set to .02 as recommended and we use the "coef" function to do see the results

```
lasso.coef<-coef(lasso,s=.02,exact = T)
lasso.coef
## 4 x 1 sparse Matrix of class "dgCMatrix"
##                          1
## (Intercept)   8.621877598
## IQ            2.443756554
## GS           -0.009174219
## SES           0.136837479
```

Results indicate that when lambda is set to .02, for a 1 unit increase in IQ there is a 2.41 point increase in language score. When GS (class size) goes up 1 unit there is a .009 point decrease in language score. Finally, when SES (socioeconomic status) increase 1 unit language score improves .13 point.

Figure 4.3 shows us the deviance explained on the x-axis. On the y-axis are the coefficients of the predictor variables. Across the top is the number of variables in the model. Below is the code

```
plot(lasso,xvar='dev',label=T)
```

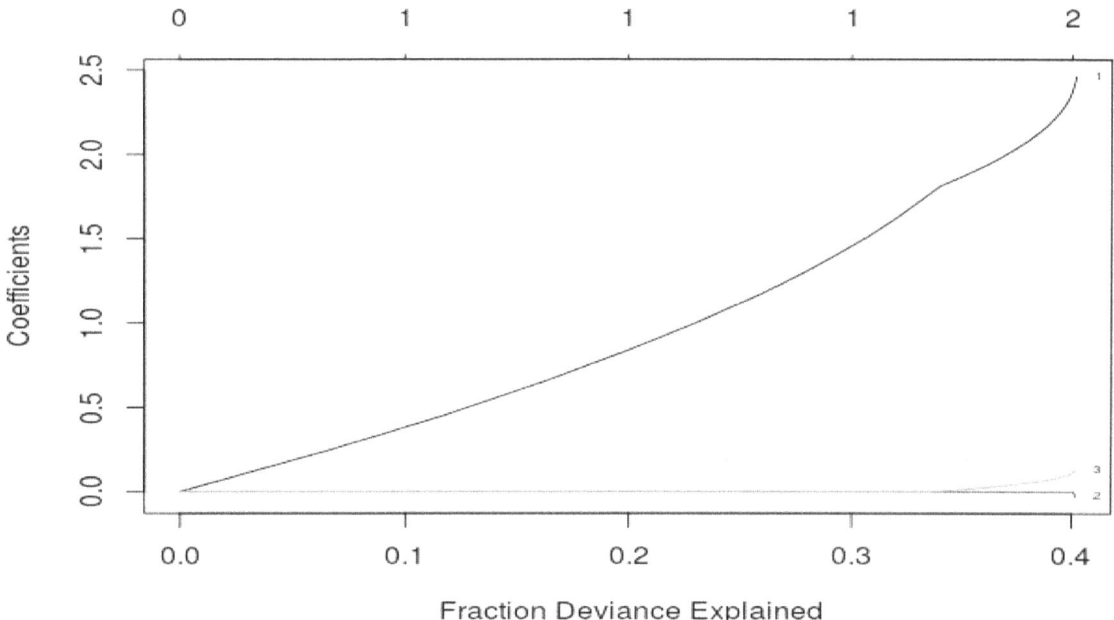

Figure 4.3: Deviance explained

If you read chapter 4, you know that If you look carefully, you can see that figure 4.3 and 4.2 are completely opposite to each other. Increasing lambda cause a decrease in the coefficients. Furthermore, increasing the fraction of deviance explained leads to an increase in the coefficient. You may remember seeing this when we used the "print"" function. As lambda became smaller, there was an increase in the deviance explained.

Model Testing

Now, we will assess our model using the test data. We need to convert the test dataset to a matrix. Then we will use the "predict"" function while setting our lambda to .02 using the 's' argument. Lastly, we will plot the results. Below is the code. Figure 4.4 is the plot of the actual and predicted results.

```
test.matrix<-as.matrix(test[,2:4])
lasso.y<-predict(lasso,newx = test.matrix,type = 'response',s=.02)
plot(lasso.y,test$lang)
```

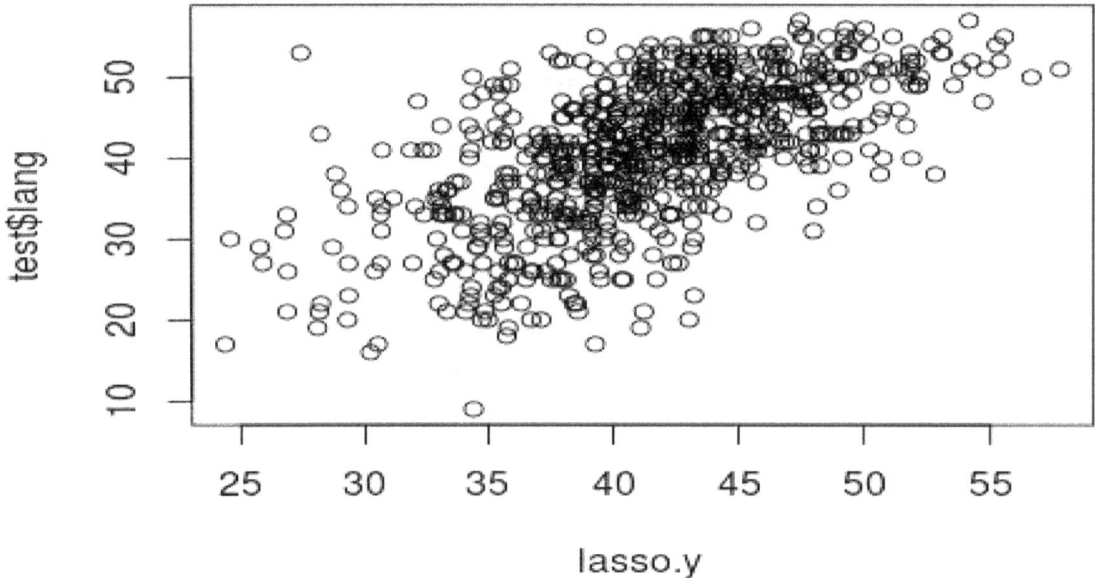

Figure 4.4: Predicted vs. actual results

The visually looks promising but as you know it does not provide enough information by its self. The finally thing we need to do is calculated the mean squared error. By its self, this number does not mean much. However, it provides a benchmark for comparing our current model with any other models that we may develop. Below is the code.

```
lasso.resid<-lasso.y-test$lang
mean(lasso.resid^2)
## [1] 50.56795
```

Knowing this number, we can, if we wanted to develop other models using other methods of analysis to try to reduce it. Generally, the lower the error the better while keeping in minds the complexity of the model.

Conclusion

In this chapter, we developed a statistical model using lasso regression. Lasso regression allows for the removal of variables through the application of a shrinkage penalty called lambda. Lasso is different from ridge in that it removes variables from the model and has an alpha set to 1. The strength of lasso is also its weakness as the ability to remove variables can impair theoretical assumptions the researcher had. Nevertheless, if you are faced with a huge number of variables lasso can provide one way of developing a model in an efficient manner.

Chapter Five:
Elastic Net Regression

Elastic net is the best of ridge and Lasso without the weaknesses of either. It combines the removal of variables like Lasso does and Ridge does not while also group variables like Ridge does but Lasso does not.

This is done by including a second tuning parameter called "alpha". Recall, that if alpha is set to 0, it is the same as ridge regression and if alpha is set to 1, it is the same as lasso regression. What this means is that elastic net regression happens whenever the alpha is not 0 or 1. If you are able to appreciate it, below is the formula used for elastic net regression.

$$(RSS + l[(1 - alpha)(S|normalized\ coefficients|2)/$$
$$2 + alpha(S|normalized\ coefficients|)]) / N)$$

As such, when working with elastic net you have to set two different tuning parameters (alpha and lambda) in order to develop a model. Therefore, our goals for this chapter are as follows.

Chapter Objectives
- Prepare a dataset for elastic net regression analysis.
- Develop a model
- Test a model
- Cross-validate a model

Data Preparation

We will go through an example of the use of elastic net using the "VietnamI" dataset from the "Ecdat" package. We want to predict how m any days a person is ill based on the other variables in the dataset. Below is some initial code for our analysis.

```
library(Ecdat);library(corrplot);library(caret);library(glmnet)
data("VietNamI")
str(VietNamI)
```

```
## 'data.frame':    27765 obs. of  12 variables:
## $ pharvis  : num  0 0 0 1 1 0 0 0 2 3 ...
## $ lnhhexp  : num  2.73 2.74 2.27 2.39 3.11 ...
## $ age      : num  3.76 2.94 2.56 3.64 3.3 ...
## $ sex      : Factor w/ 2 levels "female","male": 2 1 2 1 2 2 1 2 1 2 ...
## $ married  : num  1 0 0 1 1 1 1 0 1 1 ...
## $ educ     : num  2 0 4 3 3 9 2 5 2 0 ...
## $ illness  : num  1 1 0 1 1 0 0 0 2 1 ...
## $ injury   : num  0 0 0 0 0 0 0 0 0 0 ...
## $ illdays  : num  7 4 0 3 10 0 0 0 4 7 ...
## $ actdays  : num  0 0 0 0 0 0 0 0 0 0 ...
## $ insurance: num  0 0 1 1 0 1 1 1 0 0 ...
## $ commune  : num  192 167 76 123 148 20 40 57 49 170 ...
## - attr(*, "na.action")=Class 'omit'  Named int 27734
##   .. ..- attr(*, "names")= chr "27734"
```

The data preparation is a little more complicated compared to other chapters. We need to check correlations like always as well as developing our training and testing sets. What is new is that we need to create what is called a grid to develop our models. We also need to create an object to handle the resampling.

Let's begin by checking the correlations among the variables. We need to exclude the "sex" variable, as it is categorical. Figure 5.1 provides us with the correlations. The code for this is below.

```
p.cor<-cor(VietNamI[,-4])
corrplot.mixed(p.cor)
```

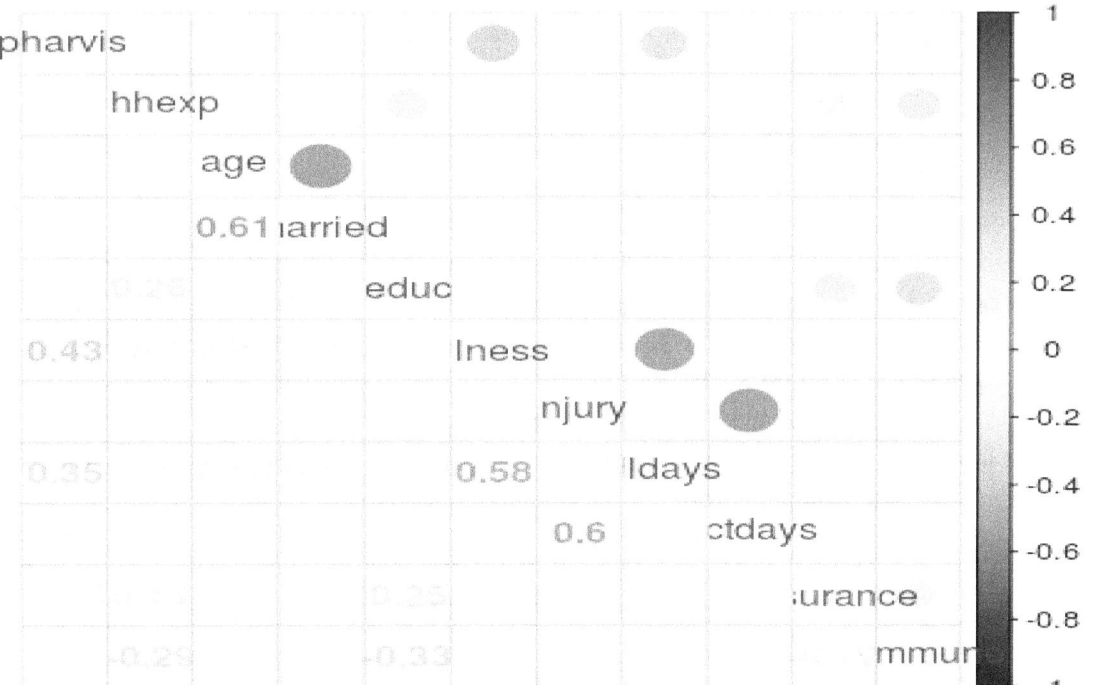

Figure 5.1: Correlations

It appears that there are no major problems with correlations. Next, we setup our training and testing datasets. We need to remove the variable "commune" because it adds no value to our results. In addition, to reduce the computational time we will only use the first 1000 rows form the data set.

```
VietNamI$commune<-NULL
VietNamI_reduced<-VietNamI[1:1000,]
ind<-sample(2,nrow(VietNamI_reduced),replace=T,prob = c(0.7,0.3))
train<-VietNamI_reduced[ind==1,]
test<-VietNamI_reduced[ind==2,]
```

Grid

We need to create a grid that will allow us to investigate different models with different combinations of alpha and lambda. This is done using the "expand.grid" function. In combination with the "seq" function below is the code.

```
grid<-expand.grid(.alpha=seq(0,1,by=.5),.lambda=seq(0,0.2,by=.1))
```

What the code above has done is create a grid that has an alpha of 0, .5, and 1 along with a lambda of 0, .1, and .2. Each combination of alpha and lambda will be analyzed. We could have made a bigger grid but elastic net is computationally heavy. If this is not clear, it will make more sense when we see the results of the analysis.

Resampling

With the grid completed, we also need to set the resampling method, which allows us to assess the validity of our model. This is done using the "trainControl" function" from the "caret" package. In the code below, "LOOCV" stands for "leave one out cross validation", which is one method of cross-validation.. This method involves using n-1 of the dataset and testing it on the one example that was left out.

```
control<-trainControl(method = "LOOCV")
```

Model Development

We are now ready to develop our model. The code is mostly self-explanatory. However, notice how our 'grid' and 'control' variables are included under the arguments 'tunerGrid' and 'trControl'. Remember that the grid is for creating several different models with different combinations of alpha and lambda while the 'control' object is for the cross-validation. We discuss cross-validation in detail later in this book.

This initial model will help us to determine the appropriate values for the alpha and lambda parameters for our final model. The code is below.

```
enet.train<-
train(illdays~.,train,method="glmnet",trControl=control,tuneGrid=grid)
enet.train
## glmnet
##
```

```
## 685 samples
##  10 predictors
##
## No pre-processing
## Resampling: Leave-One-Out Cross-Validation
## Summary of sample sizes: 684, 684, 684, 684, 684, 684, ...
## Resampling results across tuning parameters:
##
##    alpha  lambda  RMSE      Rsquared
##    0.0    0.0     4.603127  0.4137374
##    0.0    0.1     4.603127  0.4137374
##    0.0    0.2     4.603127  0.4137374
##    0.5    0.0     4.605580  0.4128863
##    0.5    0.1     4.601783  0.4138154
##    0.5    0.2     4.612353  0.4117662
##    1.0    0.0     4.605478  0.4129156
##    1.0    0.1     4.612810  0.4112516
##    1.0    0.2     4.619336  0.4113584
##
## RMSE was used to select the optimal model using  the smallest value.
## The final values used for the model were alpha = 0.5 and lambda = 0.1.
```

The output lists all the possible alpha and lambda values that were in the 'grid' that we created. In addition, the output even tells us which combination of alpha and lambda was the best. For our purposes, the alpha will be .5 and the lambda .2. Also, notice that the r-square is also included.

We will set our model and run it on the test set. Remember there are several things that the "glmnet" function does not like.

1. Factor variables must be changed to dummy variables
2. Dataframes must be changed to matrices
3. The matrices must have the predictor variables in one matrix and the outcome variable in a separate matrix.

We need to convert the "sex" variable to a dummy variable for the "glmnet" function. We next have to make matrices for the predictor variables and for our outcome variable "illdays". After doing this, we will create our model with the lambda and alpha set to the recommendations we learned previously. All the code below was used in prior chapters.

```
train$sex<-model.matrix( ~ sex - 1, data=train ) #convert to dummy variable
test$sex<-model.matrix( ~ sex - 1, data=test )
predictor_variables<-as.matrix(train[,-9]) #-9 removes the dependent variable
days_ill<-as.matrix(train$illdays)
enet<-glmnet(predictor_variables,days_ill,family = "gaussian",alpha =
0.5,lambda = .2)
```
We can now look at specific coefficients by using the "coef" function.
```
enet.coef<-coef(enet,lambda=.2,alpha=.5,exact=T)
enet.coef
```

```
## 12 x 1 sparse Matrix of class "dgCMatrix"
##                          s0
## (Intercept)    -1.125551156
## pharvis         0.579115315
## lnhhexp        -0.162156511
## age             0.780189068
## sex.sexfemale   0.040821532
## sex.sexmale    -0.002989461
## married         .
## educ            .
## illness         2.964048918
## injury          .
## actdays         0.684289251
## insurance       0.114556627
```

You can see for yourself that several variables were removed from the model. Medical expenses (lnhhexp), sex, education, injury, and insurance do not play a role in the number of days ill of an individual in Vietnam.

Model Testing

With our model developed. We now can test it using the "predict" function. However, we first need to convert our test dataframe into a matrix and remove the outcome variable from it. Then we set our model with the appropriate alpha and lambda values.

```
test.matrix<-as.matrix(test[,-9]) #-9 removes the dependent variable
enet.y<-predict(enet, newx = test.matrix, type = "response",
lambda=.2,alpha=.5)
```

Let's plot our results. They are below in figure 5.3

```
plot(enet.y)
```

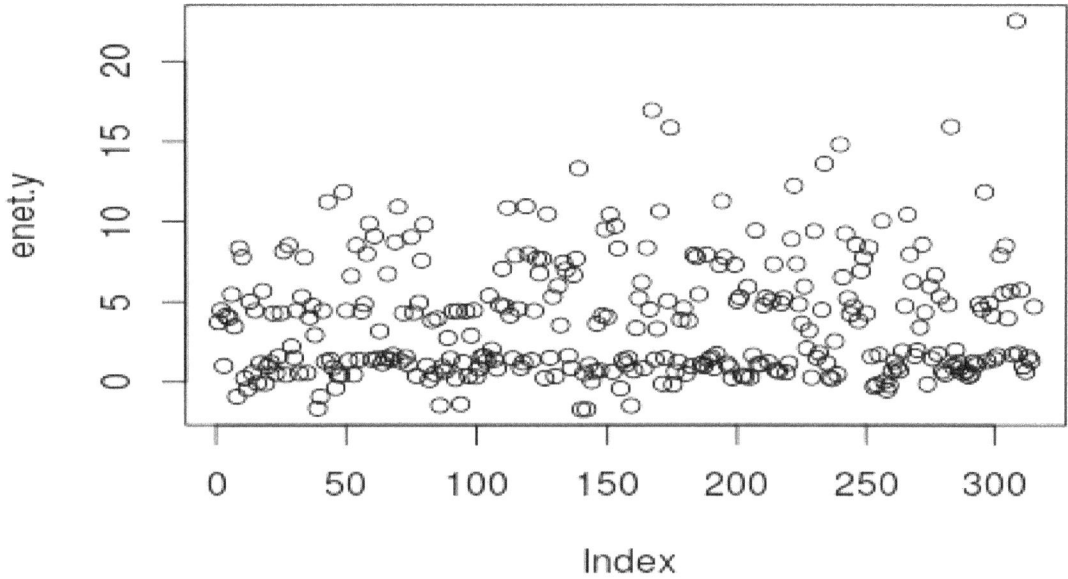

Figure 5.3: Residual vs. predicted results

The visual of our results does not look promising. There appears to be little relationship between the two. Let's check the mean squared error (MSE)

```
enet.resid<-enet.y-test$illdays
mean(enet.resid^2)
## [1] 30.40465
```

Remember the MSE means nothing unless it is compared to another model

Cross-Validation

We will now do a cross-validation of our model. We will use the k-folds. This type of cross-validation the data is partition into k sets and a separate model is create for k − 1 sets. The results are then average.

To do this, we need to set the seed and then use the "cv.glmnet" function to develop the cross validated model. The code is a little different because we have to have more than one value of lambda in order to make the model. We do this by inserting multiple values of lambda using the 'c' function. To have values of lambda that are decimals we divide them by 20 or any number you want. The code also uses the plot function to make a plot. Below are the results in figure 5.4.

```
set.seed(317)
enet.cv<-cv.glmnet(predictor_variables,days_ill, alpha=.5,lambda = c((1:100)
/ 20))
plot(enet.cv)
```

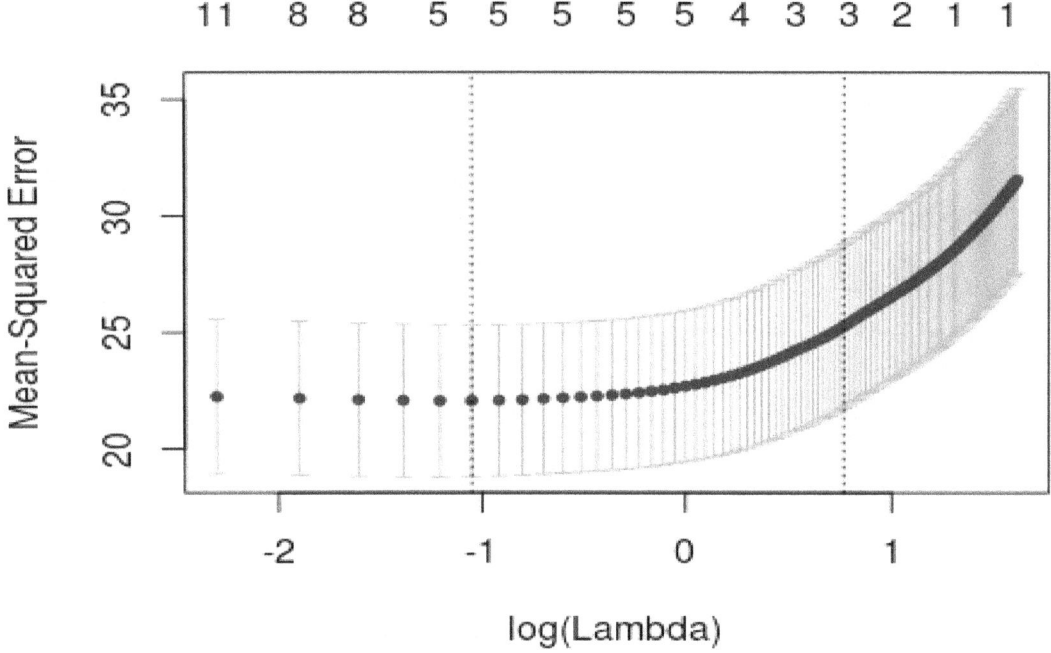

Figure 5.4: Cross-validated model

You can see that as the number of features is reduced (see the numbers on the top of the plot) the MSE increases (y-axis). In addition, as the lambda increases there is also an increase in the MSE but only when the number of variables is reduced as well.

The dotted vertical lines in the plot represent the minimum MSE for a set lambda (on the left) and the one standard error from the minimum (on the right). You can extract these two lambda values using the code below.

```
enet.cv$lambda.min
## [1] 0.35
enet.cv$lambda.1se
## [1] 2.15
```

We can see the coefficients for a lambda that is one standard error away by using the code below. This will give us an alternative idea for what to set the model parameters to when we want to predict.

```
coef(enet.cv,s="lambda.1se")

## 12 x 1 sparse Matrix of class "dgCMatrix"
##                          1
## (Intercept)    1.4570370
## pharvis        0.1505544
## lnhhexp        .
## age            0.1022935
## sex.sexfemale  .
## sex.sexmale    .
## married        .
## educ           .
## illness        2.2814262
## injury         .
## actdays        .
## insurance      .
```

Using the one standard error lambda, we lose most of our features. We can now see if the model improves by rerunning it with this information.

```
enet.y.cv<-predict(enet.cv,newx =
test.matrix,type='response',lambda="lambda.1se",alpha=.5)
enet.cv.resid<-enet.y.cv-test$illdays
mean(enet.cv.resid^2)
## [1] 31.59872
```

Well, there has been a slight increase in the MSE. Therefore, our cross-validated model is slightly worst than the original model but not by much. Another important point is that the cross-validation we did with this model can be done with the other models in the prior chapters with slight modifications to the code

Conclusion

Elastic net provides another approach to conducting regression analysis. Elastic net provides the best of ridge and lasso regression. This gives the impression that it is perfect for regression. However, it can be computationally demanding as you try to create several models with different lambdas and alphas. However, as computing power improves this should not continue to be a problem.

Chapter Six: Linear Discriminant Analysis

Discriminant analysis is similar to linear regression with the exception that the dependent variable in discriminant analysis is categorical instead of continuous as in linear regression. This means that discriminant analysis classifies examples in the test set based on the predictor variables rather than predict numeric values as in linear regression.

The classification in discriminant analysis utilizes probability the details of which are beyond the scope of this book. There are some problems with discriminant analysis. For example, it only works well if there is clear separation between the categories in the dependent variable. If the distinction between the categories is muddy then logistic regression is probably an alternative choice. Furthermore, there is a need for multivariate normal distribution.

In this chapter, we will do the following...

Chapter Objectives
- Prepare data for linear discriminant analysis
- Develop a model
- Test a model

Data Preparation
We will use the "Star" dataset from the "Ecdat" package. What we will do is try to predict the type of class the students study in (regular, small, or regular with aide) using their math scores, reading scores, and the teaching experience of the teacher. Below is the initial code.

```
library(Ecdat)
library(MASS)
data(Star)
```

We first need to examine the data by using the "str" function

```
str(Star)
## 'data.frame':    5748 obs. of  8 variables:
##  $ tmathssk: int  473 536 463 559 489 454 423 500 439 528 ...
##  $ treadssk: int  447 450 439 448 447 431 395 451 478 455 ...
##  $ classk  : Factor w/ 3 levels "regular","small.class",..: 2 2 3 1 2 1 3
1 2 2 ...
##  $ totexpk : int  7 21 0 16 5 8 17 3 11 10 ...
##  $ sex     : Factor w/ 2 levels "girl","boy": 1 1 2 2 2 2 1 1 1 1 ...
##  $ freelunk: Factor w/ 2 levels "no","yes": 1 1 2 1 2 2 2 1 1 1 ...
##  $ race    : Factor w/ 3 levels "white","black",..: 1 2 2 1 1 1 2 1 2 1
...
##  $ schidkn : int  63 20 19 69 79 5 16 56 11 66 ...
##  - attr(*, "na.action")=Class 'omit'  Named int [1:5850] 1 4 6 7 8 9 10 15
16 17 ...
##   .. ..- attr(*, "names")= chr [1:5850] "1" "4" "6" "7" ...
```

We will use the following variables
- dependent variable = classk (class type)
- independent variable = tmathssk (Math score)
- independent variable = treadssk (Reading score)
- independent variable = totexpk (Teaching experience)

We will now examine the data visually by looking at histograms for our independent variables and a table for our dependent variable. Figure 6.1 is a histogram of the math scores.

```
hist(Star$tmathssk)
```

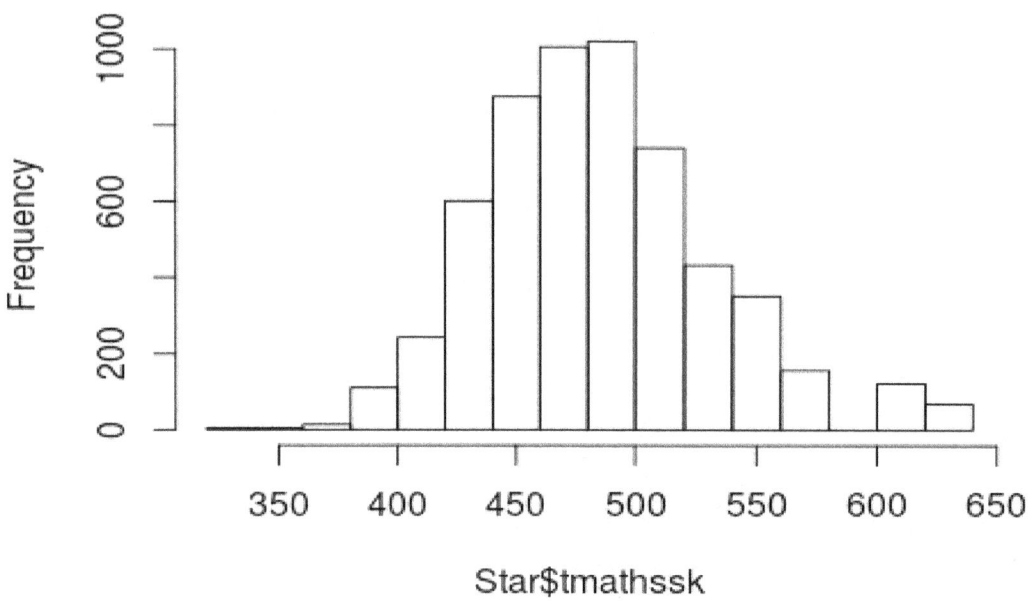

Figure 6.1: Histogram of math scores

Math scores look normally distributed. Therefore, we will now examine the distribution of the reading scores in figure 6.2.

```
hist(Star$treadssk)
```

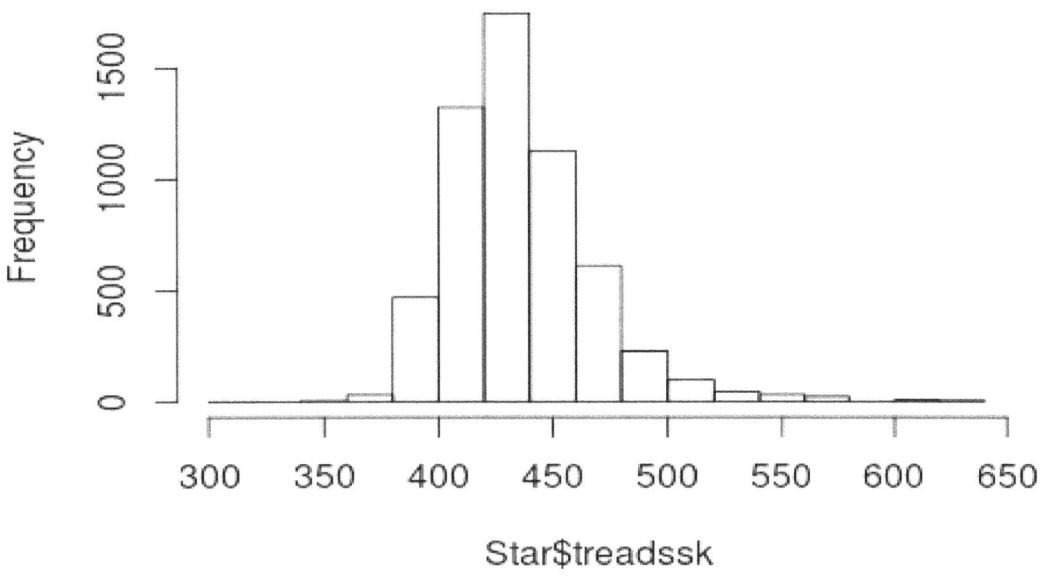

Figure 6.2: Histogram of reading scores

Reading scores also looks okay. We will now look at the histogram of teaching experience in figure 6.3

```
hist(Star$totexpk)
```

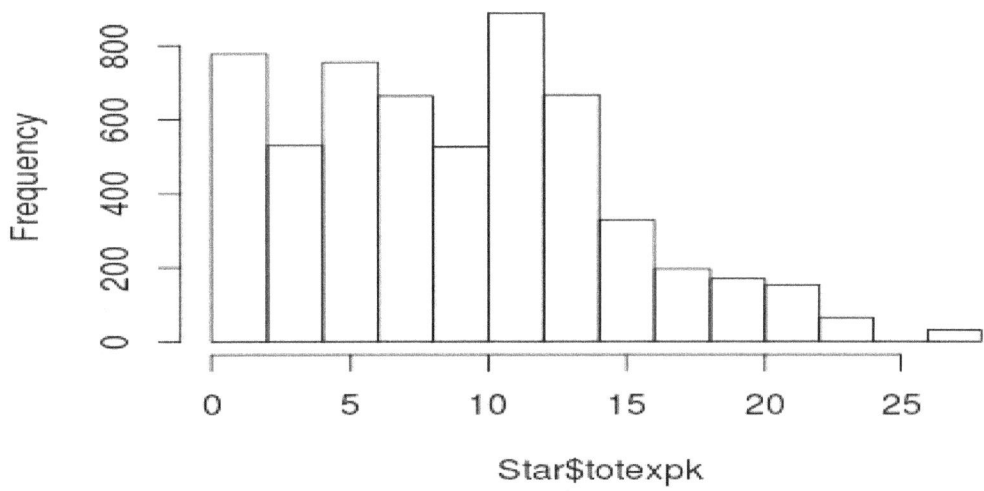

Figure 6.3: Histogram of teaching experience

There is clearly a problem here, as the data does not appear to be normally distributed at all. To deal with this problem we need to perform a transformation. We will try a square root transformation to see if this helps to alleviate the problem. Below is the code for this analysis followed by figure 6.4, which is the transformed variable of teaching experience .

```
star.sqrt<-Star
star.sqrt$totexpk.sqrt<-sqrt(star.sqrt$totexpk)
hist(sqrt(star.sqrt$totexpk))
```

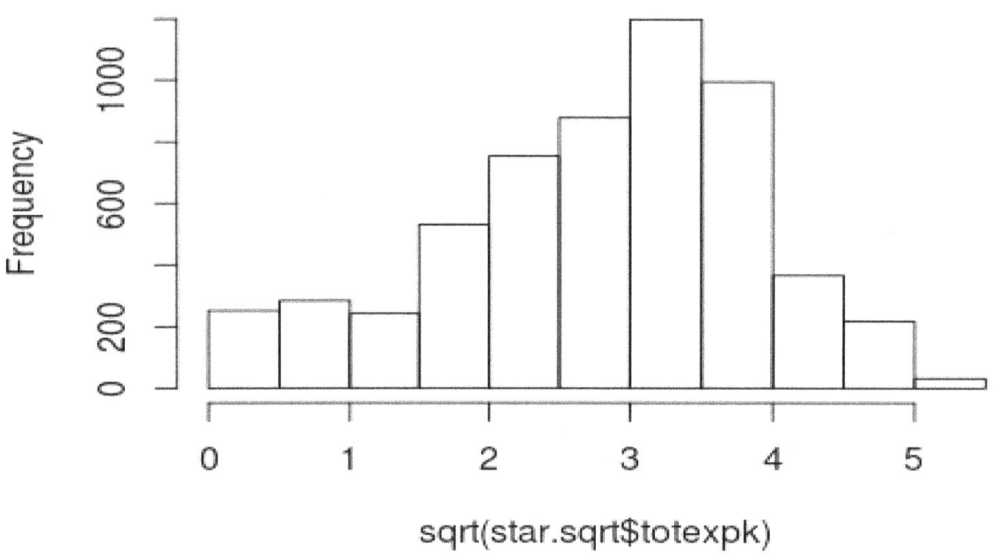

Figure 6.4: Square root transformation of teaching experience

As you can see, teaching experience looks much better. We know need to examine more closely the dependent variable of class type. We need to see what the proportions are for the three categories within this variable. The two functions involved in this are the "table" function and the "prop.table" function. Below is the code.

```
prop.table(table(Star$classk))
##
##          regular        small.class regular.with.aide
##        0.3479471          0.3014962         0.3505567
```

This information will be useful when we need to set the prior probabilities for our model. It looks as if there is a 1 in 3 chance that a student is in one of the three possible categories of the dependent variable. We now need to check the correlation among the independent variables we will use. Below is the code. Figure 6.5 is the correlation plot.

```
cor.star<-
data.frame(star.sqrt$tmathssk,star.sqrt$treadssk,star.sqrt$totexpk.sqrt)
keep<-cor(cor.star)
corrplot(cor.star)
```

Figure 6.5: Correlation plot

None of the correlations are too bad. When conducting a discriminant analysis it is important that all of the continuous variables are of the same scale. Therefore, we will scale all of the independent variables. After this ,we will divide our dataset into training and testing sets. Below is the code.

```
star.sqrt$tmathssk<-scale(star.sqrt$tmathssk)
star.sqrt$treadssk<-scale(star.sqrt$treadssk)
star.sqrt$totexpk.sqrt<-scale(star.sqrt$totexpk.sqrt)
ind=sample(2,nrow(star.sqrt),replace=T,prob=c(.7,.3))
train.star<-star.sqrt[ind==1,]
test.star<-star.sqrt[ind==2,]
```

Model Development

We can now develop our model using linear discriminant analysis. Below is the code. The code is the same as that for a regression model. The only thing new is the use of the 'prior' argument, which has been set to 1/3 for each group in the dependent variable. We are using the number 1/3 because these were the proportions we found when we used the 'prop.table' function earlier. The code is below followed by the results.

```
train.lda<-lda(classk~tmathssk+treadssk+totexpk.sqrt, data =
train.star,prior=c(1,1,1)/3)
train.lda
## Call:
## lda(classk ~ tmathssk + treadssk + totexpk.sqrt, data = train.star,
##      prior = c(1, 1, 1)/3)
##
## Prior probabilities of groups:
##          regular        small.class regular.with.aide
##        0.3333333          0.3333333         0.3333333
##
## Group means:
##                       tmathssk    treadssk totexpk.sqrt
## regular             -0.05120918 -0.08276654  -0.05417798
## small.class          0.12250726  0.11683088  -0.04465934
## regular.with.aide   -0.04087544 -0.03992069   0.08970326
##
## Coefficients of linear discriminants:
##                        LD1         LD2
## tmathssk         0.4356699 -0.1929310
## treadssk         0.5801416  0.6013748
## totexpk.sqrt    -0.5002585  0.8396815
##
## Proportion of trace:
##    LD1    LD2
## 0.734 0.266
```

The information at the top is the code used for the model. Next, are the group probabilities that we set using the 'prior' argument in the code. After this, comes the group means. This information tells you the means for each group for each independent variable and is used when assigning an example to one of the groups.

The coefficients of linear discriminant are the same as regression coefficients. The difference being that if you actually use them in an equation it will give you the probability that a specific example is in one group or another. Notice how we have two discriminant functions LD1 and LD2. This is normal, as the maximum number of functions a model can make is the number of independent variables minus 1. Since we have 3 independent variables 3 minus 1 is 2.

Lastly, the proportion of trace is similar to results in principal component analysis and tells you the proportion of variance explained by the discriminant functions. Ld1 explains 73%

while LD2 explains 27% of the variance. The discriminant functions are the actual equations used to develop the boundaries separating the groups from each other.

Model Testing

Now we will take the trained model and see how it does with the test set. We create a new model called "predict.lda" and use our "train.lda" model and the test data called "test.star"

```
predict.lda<-predict(train.lda,newdata = test.star)
```

We can use the "table" function to see how well our model has done. We can do this because we actual know what class our data is beforehand. What we need to do is compare this to what our model predicted. Therefore, we compare "classk" of our "test.star" dataset with the "class" predicted by the "predict.lda" model.

```
table(test.star$classk,predict.lda$class)
##
##                      regular small.class regular.with.aide
##    regular              156         203               238
##    small.class          132         211               173
##    regular.with.aide    184         178               271
```

The results are pretty bad. For example, in the first row called "regular" We have 156 examples that were classified as "regular" and predicted as "regular". 203 examples were classified as "regular" but predicted as "small.class", etc. To find out how well are model did you add together the examples across the diagonal from left to right and divide by the total number of examples. Below is the code

```
(156+211+271)/1746
## [1] 0.36504
```

Only 37% accurate, terrible but ok as this is fine for a demonstration of linear discriminant analysis. Since we only have two discriminant function we can plot these two dimensions with some of our examples to see how they appear visually. In figure 6.6, there is a visual of the first 50 examples classified by the predict.lda model.

```
plot(predict.lda$x[1:50])
text(predict.lda$x[1:50],as.character(predict.lda$class[1:50]),col=as.numeric
(predict.lda$class[1:50]))
abline(h=0,col="blue")
abline(v=0,col="blue")
```

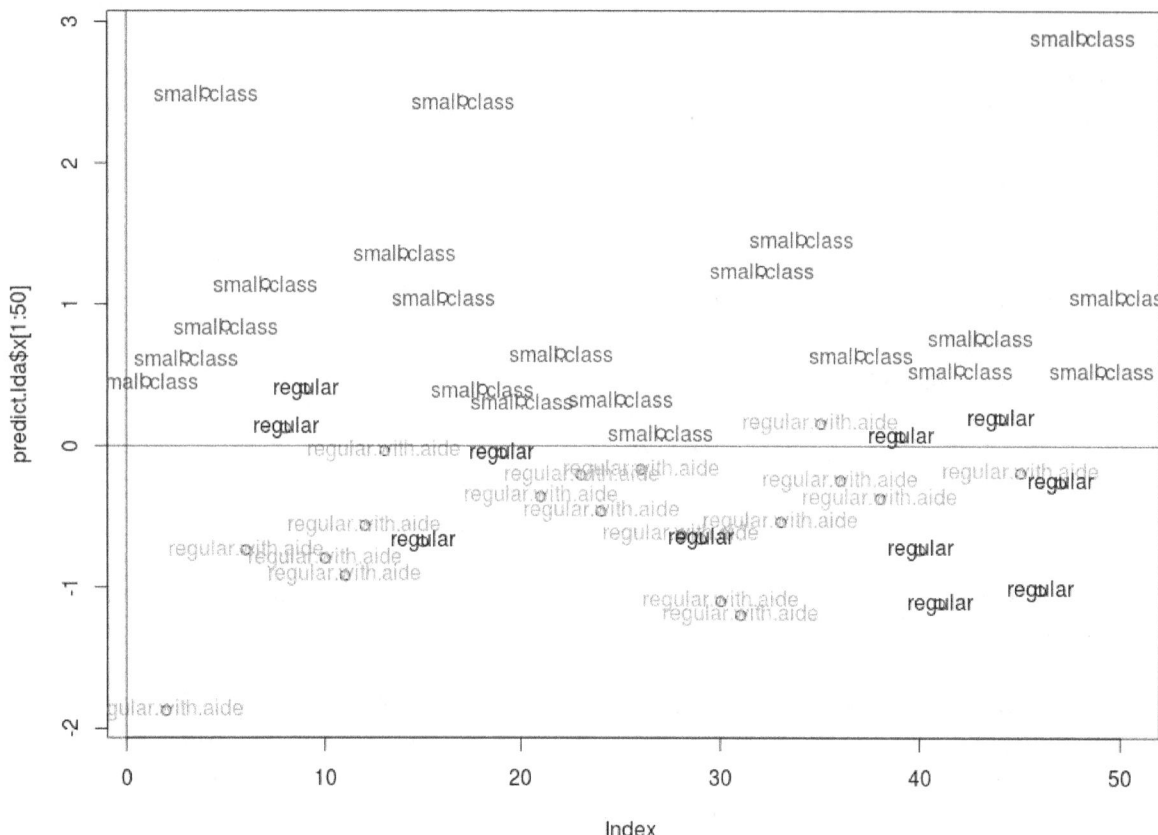

Figure 6.6: Plot of model results

The vertical line, doesn't seem to discriminant anything as it off to the side and not separating any of the data. However, the horizontal line, does a good of dividing the "regular.with.aide" from the "small.class". In order improve our model we need additional independent variables to help to distinguish the groups in the dependent variable.

Conclusion

Linear discriminant model is for the purpose of classification. Using probability it is able to predict which group an example belongs to based on the coefficients developed in the training model. A primary weakness in this type of analysis is that there needs to be a clear separation between the groups in the dependent variable. In the example, of this chapter, that was clearly not the case.

Chapter Seven: Quadratic Discriminant Analysis

In this chapter, we will conduct a quadratic discriminant analysis (QDA). If you read chapter six, you know that discriminant analysis is used when the dependent variable is categorical. LDA assumes shared covariance in the dependent variable categories will QDA allows for each category in the dependent variable to have its own variance. This allows quadratic discriminant analysis to create non-linear boundaries when separating examples.

Both LDA and QDA are used in situations in which there is a clear separation between the classes you want to predict. If the categories are fuzzier, logistic regression is often the better choice.

This chapter is slightly different in that we are going to first make a linear discriminant model and compare it with a quadratic discriminant model. The objectives of this chapter are as follows.

Chapter Objectives
- Prepare a dataset for analysis
- Develop a linear discriminant analysis model
- Assess the linear model
- Develop a quadratic discriminant analysis model
- Assess the quadratic model
- Compare the two models

Data Preparation
For our example, we will use the "Mathlevel" dataset found in the "Ecdat" package. Our goal will be to predict the sex of a respondent based on SAT math score, major, foreign language proficiency, as well as the number of math, physics, and chemistry classes a respondent took. Below is some initial code to start our analysis.

```
library(MASS);library(Ecdat) data("Mathlevel")
```

We will create a dataset called 'math' that contains the "Mathlevel" data. After this, we need to set our seed for the purpose of reproducibility using the "set.seed" function. Lastly, we will split the data using the "sample" function using a 70/30 split. The training dataset will be called "math.train" and the testing dataset will be called "math.test". Below is the code

```
set.seed(123)
math.ind<-sample(2,nrow(math),replace=T,prob = c(0.7,0.3))
math.train<-math[math.ind==1,]
math.test<-math[math.ind==2,]
```

Model Development (LDA)

Now we will make our model, we will call it "lda.math", and it will include all available variables in the "math.train" dataset. Next, we will check the results by calling the model. Finally, we will examine the plot to see how our model is doing in terms of separating the two groups in figure 7.1. Below is the code.

```
lda.math<-lda(sex~.,math.train)
lda.math
## Call:
## lda(sex ~ ., data = math.train)
##
## Prior probabilities of groups:
##      male    female
## 0.5986079 0.4013921
##
## Group means:
##        mathlevel.L mathlevel.Q mathlevel.C mathlevel^4 mathlevel^5
## male   -0.10767593  0.01141838 -0.05854724   0.2070778  0.05032544
## female -0.05571153  0.05360844 -0.08967303   0.2030860 -0.01072169
##        mathlevel^6      sat languageyes  majoreco  majoross   majorns
## male    -0.2214849 632.9457  0.07751938 0.3914729 0.1472868 0.1782946
## female  -0.2226767 613.6416  0.19653179 0.2601156 0.1907514 0.2485549
##         majorhum mathcourse physiccourse chemistcourse
## male   0.05426357   1.441860    0.7441860      1.046512
## female 0.07514451   1.421965    0.6531792      1.040462
##
## Coefficients of linear discriminants:
##                     LD1
## mathlevel.L    1.38456344
## mathlevel.Q    0.24285832
## mathlevel.C   -0.53326543
## mathlevel^4    0.11292817
## mathlevel^5   -1.24162715
## mathlevel^6   -0.06374548
## sat           -0.01043648
## languageyes    1.50558721
## majoreco      -0.54528930
## majoross       0.61129797
```

60

```
## majorns        0.41574298
## majorhum       0.33469586
## mathcourse    -0.07973960
## physiccourse  -0.53174168
## chemistcourse  0.16124610
```

Since there is only one discriminant function, we can plot this as a histogram. The plot is below in figure 7.1.

```
plot(lda.math,type='both')
```

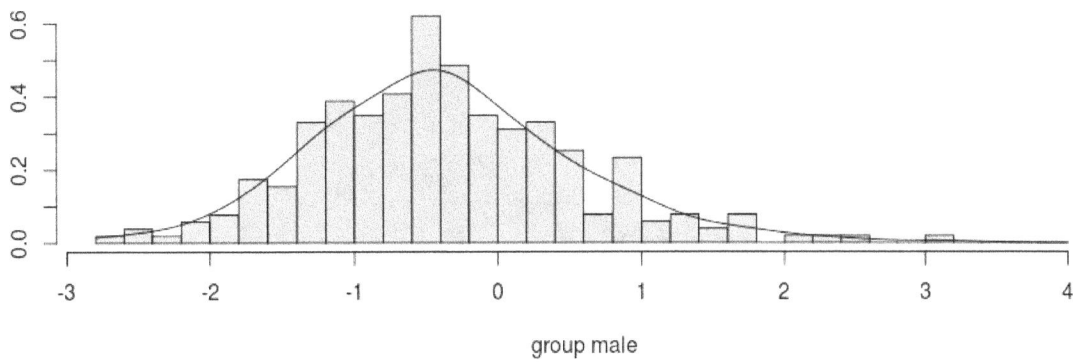

group male

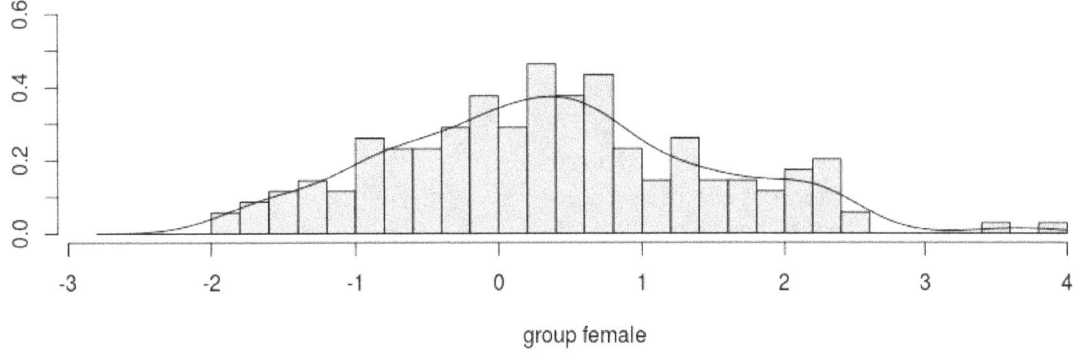

group female

Figure 7.1: Plot of linear model

Calling "lda.math" gives us the details of our model. It starts be indicating the prior probabilities of someone being male or female. Next is the means for each variable by sex. The last part is the coefficients of the linear discriminant. Each of these values is used to determining the probability that a particular example is male or female. This is similar to a regression equation.

Notice how there is only one discriminant function. Remember that it is possible for the number of discriminants = total independent variables - 1. However, fewer discriminants will be developed if additional ones do not provide any additional explanation of the variance. Because there is only one discriminant function we are able to make the one-dimensional (aka histogram) of the classification shown in figure 7.1.

The plot in figure 7.1 indicates a problem. There is a great deal of overlap between male and females in the model. What this indicates is that there is a lot of misclassification going on as the two groups are not clearly separated based on our independent variables. Furthermore, this means that logistic regression is probably a better choice for distinguishing between male and females. However, since this is for demonstrating purposes we will not worry about this.

We will now use the "predict" function on the training set data to see how well our model classifies the respondents by gender. We will then compare the prediction of the model with the actual classification. Below is the code.

```
math.lda.predict<-predict(lda.math)
math.train$lda<-math.lda.predict$class
table(math.train$lda,math.train$sex)
##
##          male female
##   male    219    100
##   female   39     73
mean(math.train$lda==math.train$sex)
## [1] 0.6774942
```

As you can see, we have a lot of misclassification happening. There is a large amount of false negatives, which indicates that there are many males being classified as female. The overall accuracy is 68%, which is good or bad depending on what your original desired target was.

Model Testing (LDA)

We will now conduct the same analysis on the test data set. Below is the code.

```
lda.math.test<-predict(lda.math,math.test)
math.test$lda<-lda.math.test$class
table(math.test$lda,math.test$sex)
##
##          male female
##   male     92     43
##   female   23     20
mean(math.test$lda==math.test$sex)
## [1] 0.6292135
```

As you can see, the results are similar. To put it simply, our model is probably not good enough for a real-world application. The main reason for this is that there is little distinction between males and females as shown in the figure 7.1. However, we can see if perhaps a quadratic discriminant analysis will do better

Quadratic Model Development and Testing

  QDA allows for each class in the dependent variable to have its own covariance rather than a shared covariance as in LDA. This allows for quadratic terms in the development of the model. To complete a QDA we need to use the "qda" function from the "MASS" package. We already have our train and test sets from the data preparation we did for the lda model so this step does not need to be repeated. Below is the code for the training data set.

```
math.qda.fit<-qda(sex~.,math.train)
math.qda.fit

## Call:
## qda(sex ~ ., data = math.train)
##
## Prior probabilities of groups:
##       male     female
## 0.5986079 0.4013921
##
## Group means:
##        mathlevel.L mathlevel.Q mathlevel.C mathlevel^4 mathlevel^5
## male   -0.10767593  0.01141838 -0.05854724   0.2070778  0.05032544
## female -0.05571153  0.05360844 -0.08967303   0.2030860 -0.01072169
##        mathlevel^6      sat languageyes  majoreco  majoross   majorns
## male    -0.2214849 632.9457  0.07751938 0.3914729 0.1472868 0.1782946
## female  -0.2226767 613.6416  0.19653179 0.2601156 0.1907514 0.2485549
##          majorhum mathcourse physiccourse chemistcourse ldafemale
## male   0.05426357   1.441860    0.7441860      1.046512 0.1511628
## female 0.07514451   1.421965    0.6531792      1.040462 0.4219653
```

  The qda model does not provide the discriminant functions. However, the means for each group and if you look closely they are the same as the means for the lda model that we developed. The means do not change; instead, the manner in which the classifications are made is what changes.

  Below is the code for the classification results for the training model.

```
math.qda.predict<-predict(math.qda.fit)
math.train$qda<-math.qda.predict$class
table(math.train$qda,math.train$sex)

##
##          male female
##   male    215     84
##   female   43     89

mean(math.train$qda==math.train$sex)

## [1] 0.7053364
```

You can see there is almost no difference. Below is the code for the test data.

```
math.qda.test<-predict(math.qda.fit,math.test)
math.test$qda<-math.qda.test$class
table(math.test$qda,math.test$sex)
##
##           male female
##   male      91     43
##   female    24     20
mean(math.test$qda==math.test$sex)
## [1] 0.6235955
```

Still disappointing, however, you now understand how to develop both linear and quadratic discriminant models. Both of these statistical tools are used for predicting categorical dependent variables.

Conclusion

Quadratic discriminant analysis allows for the boundaries that are developed around example in a model to be non-linear. This heighten flexibility can be useful when the data does not have linear characteristics.

Chapter Eight: Logistic Regression

Logistic regression like linear and quadratic discriminant analysis, involves the use of a categorical dependent variable. As such, discriminant analysis and logistic regression provide a way to predict the categories in a categorical dependent variable. The differences between these two types of analysis are how they analyze the data as well as the context in which they are used. Discriminant analysis needs data in which the categories in the dependent variable are clearly separated. In addition, discriminant analysis calculates several linear discriminants to set the boundaries for prediction.

Logistic regression can better handle data in which the separation between the groups is not clear. Furthermore, logistic regression calculates probabilities in order to calculate odds and odds ratios in order to classify examples.

Chapter Objectives
- Define key terms related to logistic regression.
- Prepare a dataset for analysis
- Develop a model employ logistic regression
- Test a model using logistic regression

We will now look at these three terms of probability, odds, and odds ratio.

Key Terms Related to Logistic Regression

Probability

Probability is probably (no pun intended) the easiest of these three terms to understand. Probability is simply the likelihood that a certain event will happen. To calculate the probability in the traditional sense you need to know the number of events and outcomes to find the probability.

Bayesian probability uses prior probabilities to develop a posterior probability based on new evidence. For example, at one point during Super Bowl LI the Atlanta Falcons had a 99.7% chance of winning. This was based on such factors as the number points they were ahead and the time remaining. As these changed, so did the probability of them winning. Yet the Patriots still found a way to win with less than a 1% chance.

Bayesian probability was also used for predicting who would win the 2016 US presidential race. It is important to remember that probability is an expression of confidence and not a guarantee as we saw in both examples.

Odds

Odds are the expression of relative probabilities. Odds are calculated using the following equation

probability of the event / 1 − probability of the event

For example, at one point during Super Bowl LI the odds of the Atlanta Falcons winning were as follows

$$0.997 / 1 − 0.997 = 332$$

This can be interpreted as the odds being 332 to 1! This means that Atlanta was 332 times more likely to win the Super Bowl then to loss the Super Bowl.

Odds are commonly used in gambling and this is probably (again no pun intended) where most of us have heard the term before. The odds are just an extension of probabilities and they are most commonly expressed as a fraction such as one in four, etc.

Odds Ratio

A ratio is the comparison of two numbers and indicates how many times one number is contained or contains another number. For example, a ratio of boys to girls is 5 to 1 it means that there are five boys for every one girl.

By extension odds ratio is the comparison of two different odds. For example, if the odds of Team A making the playoffs are 45% and the odds of Team B making the playoffs is 35% the odds ratio is calculated as follows.

$$0.45 / 0.35 = 1.28$$

Team A is 1.28 more likely to make the playoffs then Team B.

The value of the odds and the odds ratio can sometimes be the same. Below is the odds ratio of the Atlanta Falcons winning and the New Patriots winning Super Bowl LI

$$0.997 / 0.003 = 332$$

As such, there is little difference between odds and odds ratio except that odds ratio is the ratio of two odds ratio. As you can tell, there is a lot of confusion about this for the average person. However, understanding these terms is critical to the application of logistic regression.

Data Preparation

We will conduct a logistic regression analysis. In our example, we want to predict Sex (male or female) when using several continuous variables from the "survey" dataset in the "MASS" package. We will cross-validate our model using K-fold cross validation. Below is some initial code to begin the analysis.

```
library(MASS);library(bestglm);library(reshape2);library(corrplot)
?MASS::survey #explains the variables in the study
```

      The first thing we need to do is remove the independent factor variables from our dataset. The reason for this is that the function that we will use for the cross-validation does not accept factors. However, factor variables are acceptable for logistic regression. We will first use the "str" function to identify factor variables and then remove them from the dataset. We also need to remove in examples that are missing data so we use the "na.omit" function for this. Below is the code.

```
str(survey)
survey$Clap<-NULL
survey$W.Hnd<-NULL
survey$Fold<-NULL
survey$Exer<-NULL
survey$Smoke<-NULL
survey$M.I<-NULL
survey<-na.omit(survey)
```

We now need to check for collinearity using the "corrplot.mixed" function form the "corrplot" package. Figure 8.1 provides the results.

```
pc<-cor(survey[,2:5])
corrplot.mixed(pc)
corrplot.mixed(pc)
```

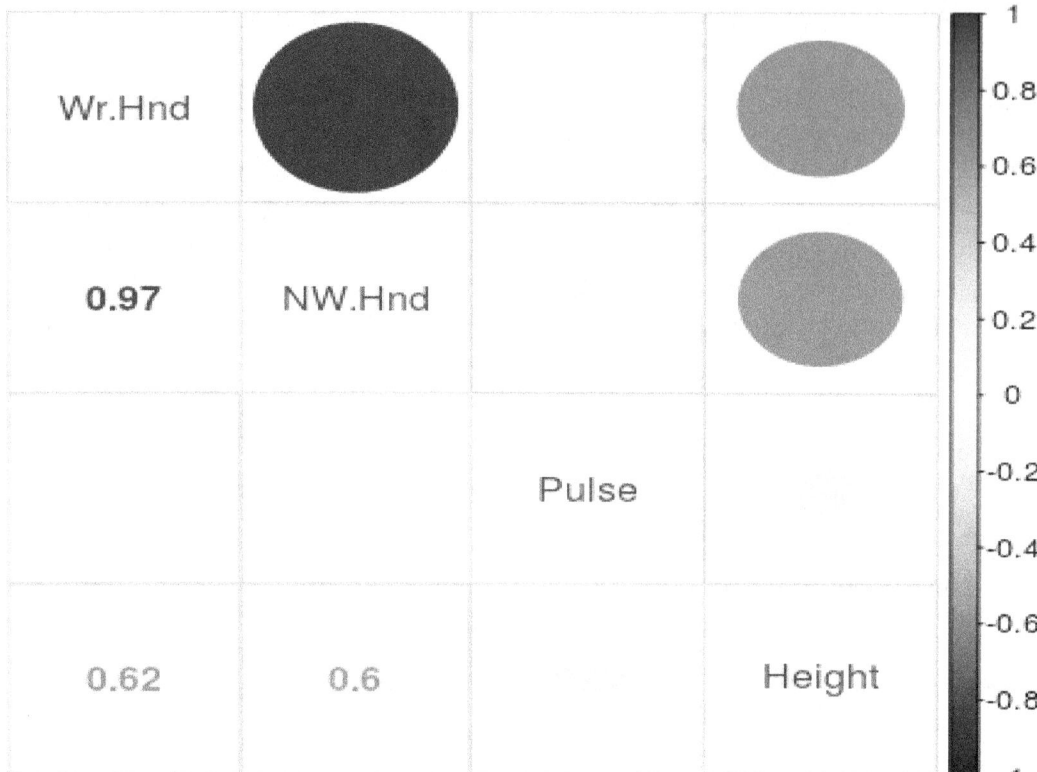

Figure 8.1: correlations

We have an extreme correlation between "Wr.Hnd" and "NW.Hnd" this makes sense because people's hands are normally the same size. Since this is a demonstration of logistic regression, we will not worry about this too much.

We now need to divide our dataset into a train and a test set. Below is the code.

```
set.seed(123)
ind<-sample(2,nrow(survey),replace=T,prob = c(0.7,0.3))
train<-survey[ind==1,]
test<-survey[ind==2,]
```

Model Development

We can now make our model. We use the "glm" function for logistic regression. We set the family argument to "binomial". Next, we look at the results as well as the odds ratios.

```
fit<-glm(Sex~.,family=binomial,train)
summary(fit)
##
## Call:
## glm(formula = Sex ~ ., family = binomial, data = train)
##
## Deviance Residuals:
##     Min       1Q   Median       3Q      Max
## -1.9875  -0.5466  -0.1395   0.3834   3.4443
##
```

```
## Coefficients:
##              Estimate Std. Error z value Pr(>|z|)
## (Intercept) -46.42175    8.74961  -5.306 1.12e-07 ***
## Wr.Hnd       -0.43499    0.66357  -0.656    0.512
## NW.Hnd        1.05633    0.70034   1.508    0.131
## Pulse        -0.02406    0.02356  -1.021    0.307
## Height        0.21062    0.05208   4.044 5.26e-05 ***
## Age           0.00894    0.05368   0.167    0.868
## ---
## Signif. codes:  0 '***' 0.001 '**' 0.01 '*' 0.05 '.' 0.1 ' ' 1
##
## (Dispersion parameter for binomial family taken to be 1)
##
##     Null deviance: 169.14  on 122  degrees of freedom
## Residual deviance:  81.15  on 117  degrees of freedom
## AIC: 93.15
##
## Number of Fisher Scoring iterations: 6
```

The results indicate that only height is useful in predicting if someone is a male or female. This makes since as men and women are usually different heights. Coefficient output is currently in log odds. For interpretation purposes, we need to exponetiated the log odds. Below is the code for this

```
exp(coef(fit))
##  (Intercept)         Wr.Hnd         NW.Hnd        Pulse        Height
## 6.907034e-21 6.472741e-01 2.875803e+00 9.762315e-01 1.234447e+00
##          Age
## 1.008980e+00
```

Exponentiated the log odds give us the odds ratios. The odds ratio tells how a one-unit increase in the independent variable leads to an increase in the odds of being male in our model. For example, for every one-unit increase in height there is a 1.23 increase in the odds of a particular example being male or a 23% increase.

To interpret exponentiated log odds you need to always subtract 1 and multiply by 100. For example, for a 1 unit increase in non-writing hand (NW.Hnd) there is a 187% increases in odds of an example being male (2.87 − 1 =1.87 *100 = 187%). You can also say that for a 1 unit increase in "NW.Hnd" a person is 1.87 times more likely to be male. Again, this makes sense as men are normally bigger than women.

We now need to see how well our model does on the train and test dataset. We first capture the probabilities and save them to the train dataset as "probs". Next, we create a "predict" variable and place the string "Female" in the same number of rows as are in the "train" dataset. Then we rewrite the "predict" variable by changing any example that has a probability above 0.5 as "Male". Then we make a table of our results to see the number correct, false positives/negatives. Lastly, we calculate the accuracy rate. Below is the code.

```
train$probs<-predict(fit, type = 'response')
train$predict<-rep('Female',123)
train$predict[train$probs>0.5]<-"Male"
table(train$predict,train$Sex)
##
##           Female Male
##    Female     61    7
##    Male        7   48
mean(train$predict==train$Sex)
## [1] 0.8861789
```

Despite the weaknesses of the model with so many insignificant variables, it is surprisingly accurate at 88.6%.

Model Testing

Let's see how well we do on the "test" dataset.

```
test$prob<-predict(fit,newdata = test, type = 'response')
test$predict<-rep('Female',46)
test$predict[test$prob>0.5]<-"Male"
table(test$predict,test$Sex)
##
##           Female Male
##    Female     17    3
##    Male        0   26
mean(test$predict==test$Sex)
## [1] 0.9347826
```

As you can see, we do even better on the test set with an accuracy of 93.4%. Our model is looking good. However, decision on improvement would be based on the use of other models of classification and or combination of features.

Conclusion

Logistic regression is another approach to classification modeling. The decision on when to use logistic regression can depend on the characteristics of the dependent variable.

Chapter Nine: Assessing Classification Model Performance

We will now look at assessing model performance much more closely. Many of the topics covered in this chapter are concepts we have been using without explaining what they are. The reason for this is that I wanted to hit the ground running by using the different techniques rather than spend several chapters in the beginning talking about all of the various machine-learning tools theoretically. Experience has shown me that people learn better from doing and then reviewing what they did rather than listening and then doing what they heard.

Specifically we will look at how to assess the performance of a classification model such as a model that uses logistic regression or linear discriminant analysis. Are primary objectives in this chapter are as follows

Chapter Objectives
- Explain the details of a confusion matrix
- Assess classification models with output from a confusion matrix
- Use visuals to assess a classification model
- Use cross-validation to assess a classification model

Confusion Matrix

We will begin by looking at a confusion matrix. In order to create a confusion matrix we need to create a model that classifies something. As such, we will use the "Wages1" dataset in the "Ecdat" package. We want to predict a person's sex based on their job experience (exper), education (school), and wage (wage) using linear discriminant analysis. We will run the code and then it will be followed with an explanation of a confusion matrix.

```
library(Ecdat);library(MASS)
data("Wages1")
```

```
class_model<-lda(sex~., data=Wages1)
class_model
## Call:
## lda(sex ~ ., data = Wages1)
##
## Prior probabilities of groups:
##    female      male
## 0.4763206 0.5236794
##
## Group means:
##           exper    school      wage
## female 7.732314 11.83748 5.146924
## male   8.326377 11.44232 6.313021
##
## Coefficients of linear discriminants:
##               LD1
## exper    0.1477352
## school  -0.3890214
## wage     0.2639368
```

The output above should be familiar from a prior chapter. We will now predict the sex of a person using the code before.

```
predict_model<-predict(class_model, Wages1)
table(predict_model$class, Wages1$sex)
##
##          female male
##   female    924  597
##   male      645 1128
```

You've seen the output in the table above but we have not really explained it explicitly. A confusion matrix is a table that is used to organize the predictions made during an analysis of data. Without making a joke, confusion matrices can be confusing, especially for those who are new to research.

The example above is a two-class confusion matrix. This matrix compares the actual class of an example with the predicted class of the model. Table 9.1 is an example of a confusion matrix.

		Predicted Class	
Actual Class		A	B
	A	1. Correctly classified as A	4. Incorrectly classified as B
	B	3. Incorrectly classified as A	2. Correctly classified as B

Table 9.1: Example confusion matrix

As you can see, sometimes things are classified correctly and sometimes they are not. If you look closely at table 9.1, each column was given a number. Below is what they represent.
1. Correctly classified as A-This means that the example was a part of the A category and the model predicted it as such. This is also known as true positives
2. Correctly classified as B-This means that the example was a part of the B category and the model predicted it as such. Also known as true negatives
3. Incorrectly classified as A-This means that the example was a part of the B category but the model predicted it to be a part of the A group. Another term for this is false negatives
4. Incorrectly classified as B-This means that the example was a part of the A category but the model predicted it to be a part of the B group. This is also known as false positives.

Table 9.2 is the results from our own R output using linear discriminant analysis (lda). Let's see if we can determine the different classifications that took place

	Predicted Class	
Actual Class	Female	Male
Female	924	597
Male	645	1128

Table 9.2: Model confusion matrix

Here is a breakdown of the results
- We have 924 true positives (females correctly classified as females)
- We have 1128 true negatives (Males correctly classified as males)
- We have 597 false positives (Females incorrectly classified as males)
- We have 645 false negatives (males incorrectly classified as females)

One of the simplest ways to compare models is to examine the confusion matrix of each model. Normally, the most accurate model is preferred if using confusion matrices is your only criteria.

The comparison of the accuracy of the confusion matrices is what we did in chapter 9 to compare the linear and quadratic discriminant models. We look at each table and calculate the accuracy using the "mean" function.

Assessing Models with Confusion Matrices Outputs

Below we will use linear and quadratic analysis to identify gender in the "Wages1" dataset. We will use the function "confusionMatrix" to uncover a lot of additional information about our models. Before explaining all the information about the function let's do the analysis. Below is the code for the lda model

```
library(Ecdat);library(MASS);library(caret)
data("Wages1")
lda_class_model<-lda(sex~., data=Wages1)
```

```
lda_predict_model<-predict(lda_class_model, Wages1)
table(lda_predict_model$class, Wages1$sex)
##
##           female male
##   female    924  597
##   male      645 1128
```

This is all the same information from the first model that we made in this chapter. Below is the new information from the "confusionMatrix" function.

```
confusionMatrix(lda_predict_model$class,Wages1$sex)
## Confusion Matrix and Statistics
##
##            Reference
## Prediction female male
##     female    924  597
##     male      645 1128
##
##                Accuracy : 0.623
##                  95% CI : (0.6061, 0.6395)
##     No Information Rate : 0.5237
##     P-Value [Acc > NIR] : <2e-16
##
##                   Kappa : 0.2432
##  Mcnemar's Test P-Value : 0.1823
##
##             Sensitivity : 0.5889
##             Specificity : 0.6539
##          Pos Pred Value : 0.6075
##          Neg Pred Value : 0.6362
##              Prevalence : 0.4763
##          Detection Rate : 0.2805
##    Detection Prevalence : 0.4617
##       Balanced Accuracy : 0.6214
##
##        'Positive' Class : female
##
```

We will explain the information above in a minute. Next, we create our qda model. The code is mostly the same except we use the "qda" function instead of the "lda" function. We also call the "qda_class_model" because this is new information.

```
qda_class_model<-qda(sex~., data=Wages1)
qda_class_model
## Call:
## qda(sex ~ ., data = Wages1)
##
```

```
## Prior probabilities of groups:
##    female      male
## 0.4763206 0.5236794
##
## Group means:
##          exper   school     wage
## female 7.732314 11.83748 5.146924
## male   8.326377 11.44232 6.313021
qda_predict_model<-predict(qda_class_model, Wages1)
table(qda_predict_model$class, Wages1$sex)
##
##         female male
##   female   1145  876
##   male      424  849
```

Next, is the "confusionMatrix" function results.

```
confusionMatrix(qda_predict_model$class,Wages1$sex)
## Confusion Matrix and Statistics
##
##           Reference
## Prediction female male
##     female   1145  876
##     male      424  849
##
##               Accuracy : 0.6053
##                 95% CI : (0.5884, 0.6221)
##    No Information Rate : 0.5237
##    P-Value [Acc > NIR] : < 2.2e-16
##
##                  Kappa : 0.2191
##  Mcnemar's Test P-Value : < 2.2e-16
##
##            Sensitivity : 0.7298
##            Specificity : 0.4922
##         Pos Pred Value : 0.5666
##         Neg Pred Value : 0.6669
##             Prevalence : 0.4763
##         Detection Rate : 0.3476
##   Detection Prevalence : 0.6135
##      Balanced Accuracy : 0.6110
##
##       'Positive' Class : female
##
```

Confusion Matrix Explanation

So what does the confusion matrix information tell us? We will look at several commonly used measures, specifically...

- Accuracy
- kappa
- error
- sensitivity
- specificity

Accuracy

Accuracy is probably the easiest statistic to understand. Accuracy is the total number of items correctly classified divided by the total number of items below is the equation

$$\text{accuracy} = \text{TP} + \text{TN} / \text{TP} + \text{TN} + \text{FP} + \text{FN}$$

TP = true positive, TN = true negative, FP = false positive, FN = false negative

In our two models, the accuracy of lda was 62% and the accuracy of qda was 61%

Accuracy can range in value from 0-1 with one representing 100% accuracy. Normally, you don't want perfect accuracy as this is an indication of overfitting and your model will probably not do well with other data.

Kappa

The kappa statistic is a measurement of accuracy of a model while taking into account chance. The closer the value is to 1 the better. The kappa for the lda model was .24 and for the qda model, it was .21. Neither is impressive

Error

Error is the opposite of accuracy and represents the percentage of examples that are incorrectly classified. This information is not in our matrices above but its equation is as follows.

$$\text{error} = \text{FP} + \text{FN} / \text{TP} + \text{TN} + \text{FP} + \text{FN}$$

Another way to see this equation is error = 1 − accuracy. The lower the error the better the model is in general. However, if error is 0 it indicates overfitting. Keep in mind that error is the inverse of accuracy. As one increases the other decreases.

Sensitivity

Sensitivity is the proportion of true positives that were correctly classified. The formula is as follows

$$\text{sensitivity} = \text{TP} / \text{TP} + \text{FN}$$

This may sound confusing but high sensitivity is useful for assessing a negative result. In other words, if I am testing people for a disease and my model has a high sensitivity. This means that the model is useful telling me a person does not have a disease. For our lda model, the sensitivity was .59 and for the qda, it was .73. For the qda model, it was much better at telling someone was not female

Specificity

Specificity measures the proportion of negative examples that were correctly classified. The formula is below

specificity = TN / TN + FP

Returning to the disease example, a high specificity is a good measure for determining if someone has a disease if he or she test positive for it. The specificity for the lda model was .65 and for the qda model, it was .56. As such, the lda was much better at determining someone was really female.

Remember that no test is foolproof and there are always false positives and negatives happening. The role of the researcher is to maximize the sensitivity or specificity based on the purpose of the model.

There are other metrics in the confusion matrix that we are not going to discuss such as Mcnemar's test, which is useful in a medical context However, there are also several metrics that the confusion matrix does not calculate, and these are listed below.

Precision

For example, Precision is the proportion of examples that are really positive. The formula is as follows

precision = TP / TP + FP

The more precise a model is the more trustworthy it is. In other words, high precision indicates that the results are relevant. For the lda model, it was .61 and for the qda model, it was .57.

Recall

Recall is a measure of the completeness of the results of a model. It is calculated as follows

recall = TP / TP + FN

This formula is the same as the formula for sensitivity. The difference is in the interpretation. High recall means that the results have a breadth to them such as in search engine results.

F-Measure

The f-measure uses recall and precision to develop another way to assess a model. The formula is below

sensitivity = 2 * TP / 2 * TP + FP + FN

The f-measure can range from 0 – 1 and is useful for comparing several potential models using one convenient number. For the lda model, it was .43 and for the qda model, it was .46.

In case this is beginning to get confusing (no pun intended) table 9.3 compares the two models so far.

	Lda model	Qda Model	Winner
Accuracy	.62	.61	Tie
Kappa	.24	.21	Lda
Error	.38	.39	Tie
Sensitivity	.59	.73	Qda
Specificity	.65	.56	Lda
Precision	.61	.57	Lda
F-measure	.43	.46	Tie

Table 9.3: Model comparison

So which is the better model? It depends on the purpose of the study. If sensitivity is most important, the lda model is best. If specificity is important, qda is the way to go. Our purpose is just to demonstrate how to obtain these different metrics.

Keep in mind that we could have also done a logistic regression model as well since logistic regression is a categorical predictor model. We could also have used other models not covered in this book such as decision trees, K nearest neighbor, etc.

Something we did not do in the example used above was create a training and testing set. I only wanted to demonstrate how this was done. However, if you create a training and testing set you need compare the prediction of the testing sets to each other. Model comparison must be done under similar circumstances otherwise they lack meaning. Since we used the whole dataset as a training set and compared the results of both models based on this dataset we are ok.

Using Visuals to Assess Classification Models

The receiver operating characteristic curve (ROC curve) is a tool used in statistical research to assess the trade-off of detecting true positives and true negatives. The origins of this tool go all the way back to WWII when engineers were trying to distinguish between true and false alarms. Now this technique is used in machine learning

Again, it is best to run an example then explain the results. We will reuse the logistic regression example from chapter 8 here. We will make three different models, which are a full model with all variables, a half-bad model with most of the variables, and bad variable with a poor predictor. Remember that the purpose was to predict the sex of a person using the variables in the dataset. Below is the initial code. This time we will have a training and test set

```
library(MASS);library(bestglm);library(reshape2);library(corrplot);library(ggplot2);library(ROCR)
data(survey)
#removed the following variables
survey$Clap<-NULL
```

```
survey$W.Hnd<-NULL
survey$Fold<-NULL
survey$Exer<-NULL
survey$Smoke<-NULL
survey$M.I<-NULL
survey<-na.omit(survey)

set.seed(123)
ind<-sample(2,nrow(survey),replace=T,prob = c(0.7,0.3))
train<-survey[ind==1,]
test<-survey[ind==2,]
fit<-glm(Sex~.,binomial,train)
exp(coef(fit))
##  (Intercept)        Wr.Hnd        NW.Hnd         Pulse        Height
## 6.907034e-21 6.472741e-01 2.875803e+00 9.762315e-01 1.234447e+00
##          Age
## 1.008980e+00
train$probs<-predict(fit, type = 'response')
train$predict<-rep('Female',123)
train$predict[train$probs>0.5]<-"Male"
table(train$predict,train$Sex)
##
##           Female Male
##    Female     61    7
##    Male        7   48
mean(train$predict==train$Sex)
## [1] 0.8861789
```

Our full model is highly accurate at almost 89%. Let's test it on the test data

```
test$prob<-predict(fit,newdata = test, type = 'response')
test$predict<-rep('Female',46)
test$predict[test$prob>0.5]<-"Male"
table(test$predict,test$Sex)
##
##           Female Male
##    Female     17    3
##    Male        0   26
mean(test$predict==test$Sex)
## [1] 0.9347826
```

Even better.

We will now create a variable called "pred.full" to begin the process of graphing the full model. Then we will use the "prediction" function. Next, we will create the "perf.full" variable to store the performance of the model. Notice, the arguments 'tpr' and 'fpr' for true positive rate and false positive rate. Lastly, we plot the results. Figure 9.1 is the ROC curve for the full model.

```
pred.full<-prediction(test$prob,test$Sex)
perf.full<-performance(pred.full,'tpr','fpr')
plot(perf.full, col=2)
```

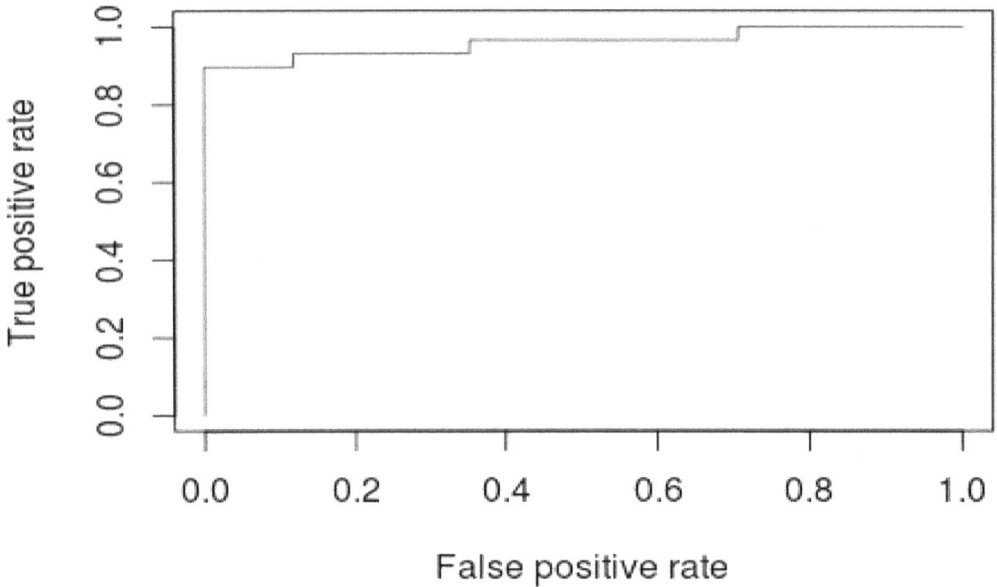

Figure 9.1:full model

In figure 9.1, on the x-axis, we have the false positive rate. As you move to the right the false positive rate increases which is bad. We want to be as close to zero as possible. On the y-axis, we have the true positive rate. Unlike the x-axis, we want the true positive rate to be as close to 100 as possible. In general, we want a low value on the x-axis and a high value on the y-axis. The more of a diagonal the line is the worst it is. The l-shape in figure 9.1 indicates that this model is reasonable good at making predictions

We are going to create our bad model now. Then we will create the ROC curve with our bad model. We will store the results of the bad model in the "test" dataset.

```
bad.fit<-glm(Sex~Age,family = binomial,test)
test$bad.probs<-predict(bad.fit,type='response')
test$bad.predict<-rep('Female',46)
test$bad.predict[test$bad.probs>0.5]='Male'
table(test$bad.predict,test$Sex)

##
##         Female Male
##   Male      17   29
```

```
mean(test$bad.predict==test$Sex)
```

```
## [1] 0.6304348
```

As you can see, the bad model is terrible. Every example was predicted as male if you look at the dataframe. The overall accuracy is 63% which is not much better than chance. Below is the code for the ROC curve and figure 9.2 is the ROC curve for the bad model

```
pred.bad<-prediction(test$bad.probs,test$Sex)
perf.bad<-performance(pred.bad,'tpr','fpr')
plot(perf.bad,col=1)
```

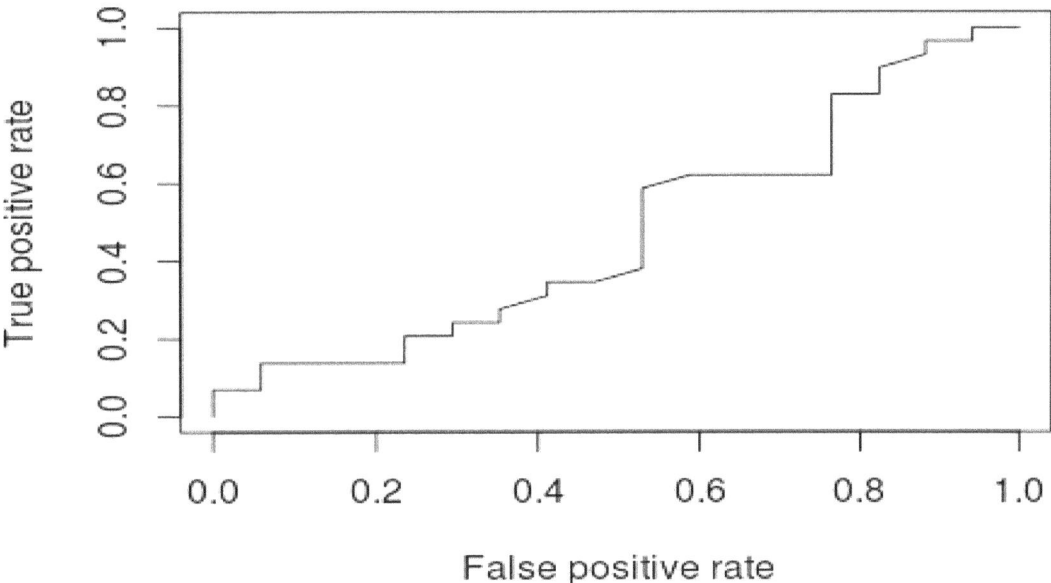

Figure 9.2: ROC curve for bad model

As, we can see the bad model is really bad. The diagonal line means that the model struggles to discern between the true and false positives.

. We repeat this process for the half-bad model. Figure 9.3 is the ROC curve for the half-bad model.

```
half.bad<-glm(Sex~Age+NW.Hnd,family = binomial,train)
```

```
train$half.bad.probs<-predict(half.bad,type='response')
train$half.bad.predict<-rep('Female',123)
train$half.bad.predict[train$half.bad.probs>0.5]='Male'
table(train$half.bad.predict,train$Sex)
```

```
##
##              Female Male
##     Female      60    13
##     Male         8    42
mean(train$half.bad.predict==train$Sex)
## [1] 0.8292683
test$half.bad.probs<-predict(half.bad,test,type = 'response')
test$half.bad.predict<-rep('Female',46)
test$half.bad.predict[test$half.bad.probs>0.5]='Male'
table(test$half.bad.predict,test$Sex)
##
##              Female Male
##     Female      17     9
##     Male         0    20
mean(test$half.bad.predict==test$Sex)
## [1] 0.8043478
pred.half.bad<-prediction(test$half.bad.probs,test$Sex)
perf.half.bad<-performance(pred.half.bad,'tpr','fpr')
plot(perf.half.bad,col=3)
```

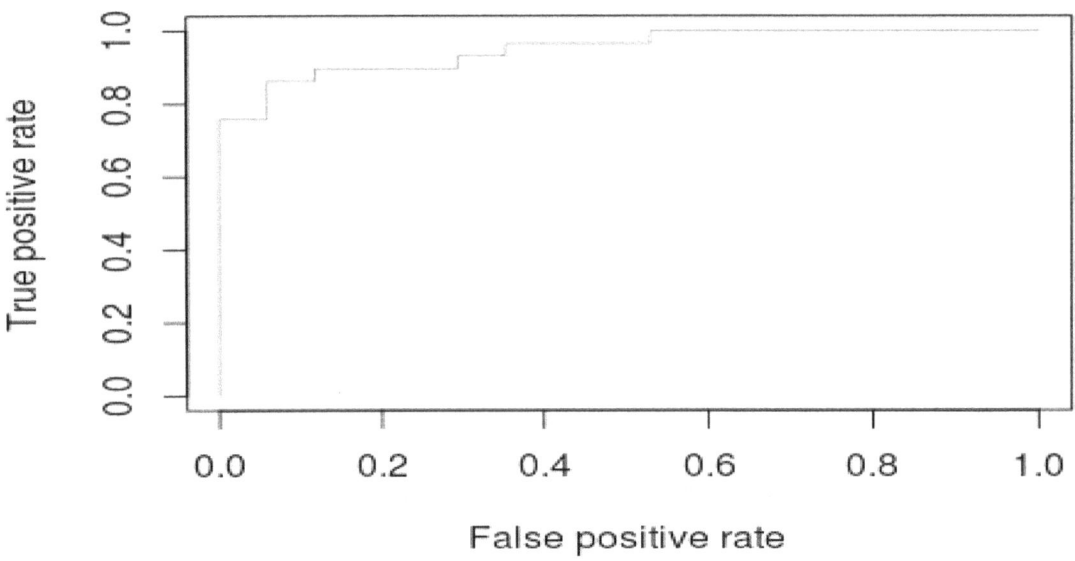

Figure 9.3: ROC curve for half-bad model

This looks better. It is not nearly as diagonal as the bad model Now let's put all the different models on one plot. Figure 9.4 shows the combined results.

```
plot(perf.bad,col=1)
plot(perf.full, col=2, add=T,lty-2)
plot(perf.half.bad,col=3,add=T,lty=3)
legend(.7,.4,c("BAD","FULL","HALF BAD",lty=1:3), 1:
```

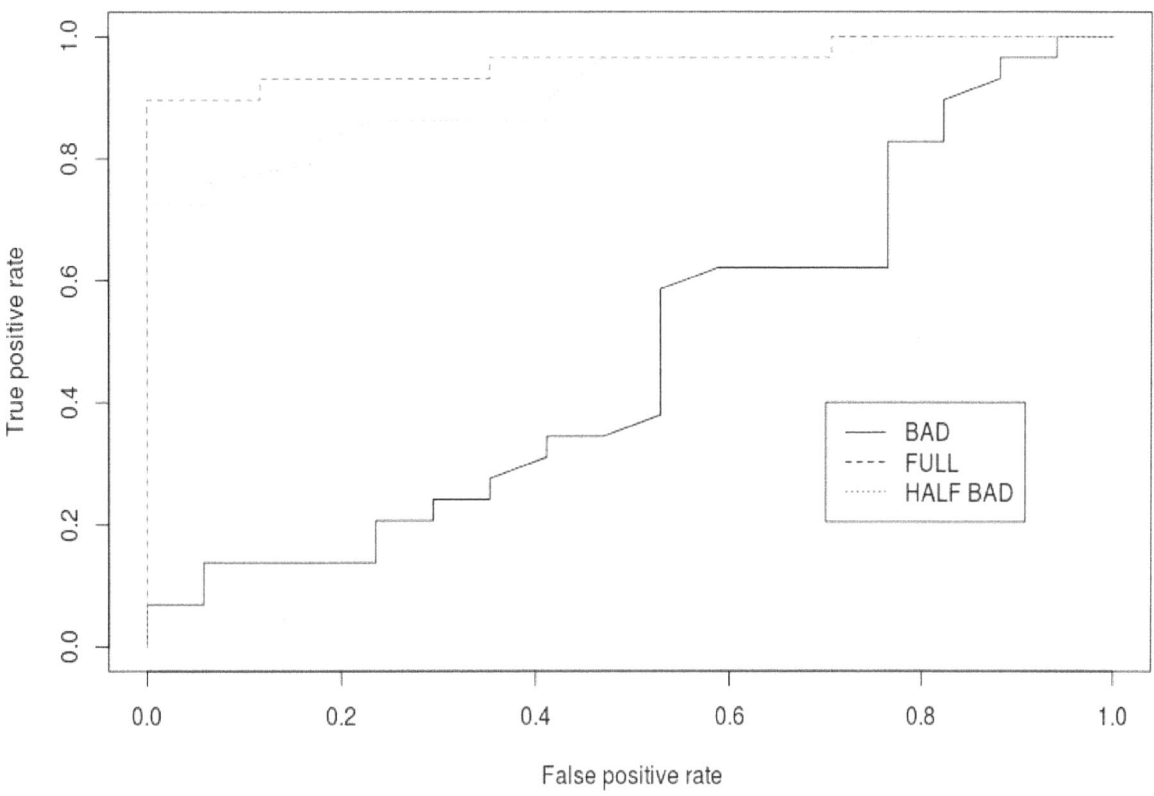

Figure 9.4: All ROC curves

As you can see, the bad model is really bad but the half bad and full model are pretty close. The last thing we will do is calculate the AUC. AUC stand for "area under the curve." The higher the number the better the model is when compared to others. Below is the code.

```
auc.bad<-performance(pred.bad,'auc')
auc.bad@y.values
## [[1]]
## [1] 0.4766734
auc.full<-performance(pred.full,"auc")
auc.full@y.values
## [[1]]
## [1] 0.959432
auc.cv<-performance(pred.half.bad,'auc')
auc.cv@y.values
## [[1]]
## [1] 0.9492901
```

The full model is superior to half-bad model and the bad model. The statistics provide support for choosing a model but they do not trump the ability of the researcher to pick a model based on factors beyond just numbers

Cross-Validation

We have been using cross-validation throughout this text without explaining what it does. Cross-validation allows you to get a sense of how your model would perform on future datasets. This is done through dividing your dataset into different folds and training the model on them. By consistently shuffling what is the testing or training set it can help to simulate collecting new data and provide insights into the models suitability for real world application

One commonly used type of cross-validation is k-fold cross-validation. K-fold cross-validation is used for determining the performance of statistical models. How it works is the data is divided into a pre-determined number of folds (called 'k'). One fold is used to test the model estimates and the other folds are used for developing it. This is done k times and the results are average based on a statistic such as the kappa to see how the model performs.

We are going to keep using the data from our first example in this text. In the "Wages1" dataset, we were trying to predict a person's sex based on experience, education, and wages. However, this time we will do a cross-validation using lda. We will need the help of the "caret" package to do this. We need to make an object in which we store some of the controls for the validation. We will create an object called "chck" for this. Below is the initial code and the code for "chck"

```
library(caret); library(Ecdat)
data(Wages1)
set.seed(1)
chck<-trainControl(method = "cv",number = 10)
```

For "chck", we used the "trainControl" function to set some criteria for the validation. The "trainControl" is a highly useful function that can be used to do much more than what we are doing here. In our code, the method is "cross-validation" and the number of folds is 10. Ten folds is the standard and normally nothing improves beyond that. Below is the code for the actually model. It is the same as before except now we use the "train" function from the "caret" package as well as add a method "lda" and the trControl argument.

```
tuned_model<-train(sex ~., data=Wages1, method="lda",trControl=chck)
tuned_model
## Linear Discriminant Analysis
##
## 3294 samples
##    3 predictors
##    2 classes: 'female', 'male'
##
## No pre-processing
## Resampling: Cross-Validated (10 fold)
## Summary of sample sizes: 2966, 2965, 2964, 2964, 2965, 2964, ...
## Resampling results:
```

```
##
##    Accuracy    Kappa
##    0.6217207  0.2407308
##
##
```

Below is the confusion matrix results.

```
predict_model<-predict(tuned_model, Wages1)
confusionMatrix(predict_model, Wages1$sex)
## Confusion Matrix and Statistics
##
##           Reference
## Prediction female male
##     female    924  597
##     male      645 1128
##
##               Accuracy : 0.623
##                 95% CI : (0.6061, 0.6395)
##    No Information Rate : 0.5237
##    P-Value [Acc > NIR] : <2e-16
##
##                  Kappa : 0.2432
##  Mcnemar's Test P-Value : 0.1823
##
##            Sensitivity : 0.5889
##            Specificity : 0.6539
##         Pos Pred Value : 0.6075
##         Neg Pred Value : 0.6362
##             Prevalence : 0.4763
##         Detection Rate : 0.2805
##   Detection Prevalence : 0.4617
##      Balanced Accuracy : 0.6214
##
##       'Positive' Class : female
##
```

Our cross-validated model is not much better than the other models we have developed with this data set (see table 9.3). However, it does provide insight into how our model might perform in the future, which is something we did not know beforehand.

Conclusion

This chapter may have been somewhat confusing. Therefore, I want to provide some key points to help to clarify things
- The main way to assess any model is by comparing it to another model
- Comparing models can be done in many ways

- o You can compare models that use different algorithms such lda and qda or lda and logistic, etc.
  - o You can compare models that use the same algorithm but have different combination of variables
  - o You can compare a model with a cross-validated version of the same model or a cross-validated model using a different algorithm
- The point is that when comparing models you need to make sure there is one difference between the two models while holding everything else constant. For example, don't change the variables in both models and at the same time change the algorithms used for analysis. This is because you will not know if the change in variables or the change in algorithm is what made the difference in the results. Modifying models requires experimentation so all the concepts of experimental design apply here as well.
- Off course, you can compare more than two models at a time. However, you need to make sure that you thoroughly track what is different about each model. In my experience, comparing multiple models starts to get confusing so it may be wiser to work with pairs of models at a time.

Chapter Ten: Assessing Numeric Model Performance

Assessing a numeric model, such as regression, is actually much simpler than a classification model. We have already done this is in several of the chapters on regression. The r-squared can be used to compare models but normally examining the error is often the strongest metric. The amount of error is measured using the mean squared error (MSE) or the rooted mean squared error (RMSE). Both of these values represent the same thing. However, you need to know this because different R packages will give you the MSE or the RMSE. The residual standard error is yet another measurement. The point is always to compare like to like otherwise you will make mistakes.

Therefore, our objectives for this chapter are as follows.

- Create a model using multiple regression
- Test the initial regression model
- Create a second regression model
- Test the second regression model
- Create a cross-validated model
- Compare the performance of the models

Initial Model Development

We are going to use the "Carseats" dataset form the "ISLR" package. We want to predict sales using all of the variables in the dataset. Below is the initial code to get things started

```
library(caret); library(ISLR);library(corrplot)

data("Carseats")
str(Carseats)

## 'data.frame':    400 obs. of  11 variables:
##  $ Sales       : num  9.5 11.22 10.06 7.4 4.15 ..
```

```
##  $ CompPrice  : num  138 111 113 117 141 124 115 136 132 132 ...
##  $ Income     : num  73 48 35 100 64 113 105 81 110 113 ...
##  $ Advertising: num  11 16 10 4 3 13 0 15 0 0 ...
##  $ Population : num  276 260 269 466 340 501 45 425 108 131 ...
##  $ Price      : num  120 83 80 97 128 72 108 120 124 124 ...
##  $ ShelveLoc  : Factor w/ 3 levels "Bad","Good","Medium": 1 2 3 3 1 1 3 2
3 3 ...
##  $ Age        : num  42 65 59 55 38 78 71 67 76 76 ...
##  $ Education  : num  17 10 12 14 13 16 15 10 10 17 ...

##  $ Urban      : Factor w/ 2 levels "No","Yes": 2 2 2 2 2 1 2 2 1 1 ...
##  $ US         : Factor w/ 2 levels "No","Yes": 2 2 2 2 1 2 1 2 1 2 ...
```

We will first check for high correlations. The code below will exclude the factor variables automatically. Figure 10.1 is the correlation matrix.

```
#cor plot
corrplot(cor(Carseats[sapply(Carseats, function(x)
!is.factor(x))]),method='number',col='black')
```

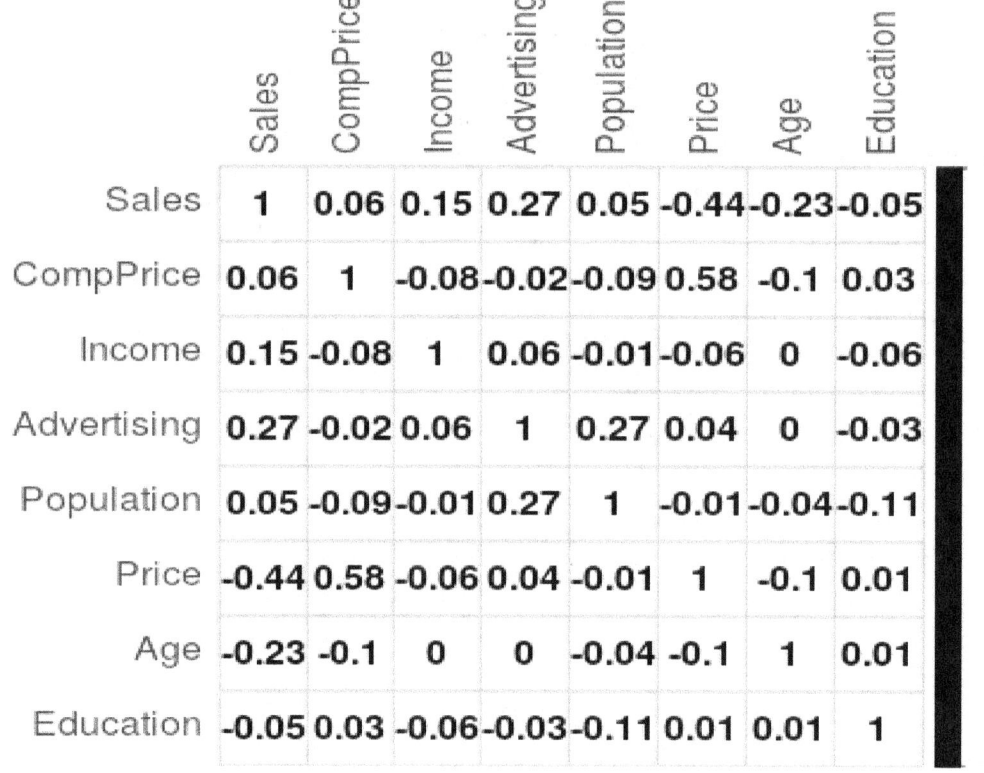

Figure 10.1: Correlational matrix

There are no major issues with correlations. We will now setup our training and testing sets and create the initial model.

```
#train and test sets
ind<-sample(2,nrow(Carseats),replace=T,prob = c(0.7,0.3))
train<-Carseats[ind==1,]
test<-Carseats[ind==2,]
initialModel<-lm(Sales~.,train)
summary(initialModel)
```

```
##
## Call:
## lm(formula = Sales ~ ., data = train)
##
## Residuals:
##     Min       1Q  Median       3Q      Max
## -2.7940 -0.7229  0.0280   0.6989   3.2485
##
## Coefficients:
##                    Estimate Std. Error t value Pr(>|t|)
## (Intercept)      5.0756194  0.7001816    7.249 4.10e-12 ***
## CompPrice        0.0940777  0.0049989   18.820  < 2e-16 ***
## Income           0.0156730  0.0021395    7.326 2.55e-12 ***
## Advertising      0.1271370  0.0130797    9.720  < 2e-16 ***
## Population       0.0002013  0.0004280    0.470    0.639
## Price           -0.0975950  0.0031444  -31.038  < 2e-16 ***
## ShelveLocGood    4.9265443  0.1843866   26.719  < 2e-16 ***
## ShelveLocMedium  2.0196194  0.1507592   13.396  < 2e-16 ***
## Age             -0.0455403  0.0037668  -12.090  < 2e-16 ***
## Education        0.0153764  0.0233008    0.660    0.510
## UrbanYes         0.1385073  0.1322158    1.048    0.296
## USYes           -0.0488057  0.1781393   -0.274    0.784
## ---
## Signif. codes:  0 '***' 0.001 '**' 0.01 '*' 0.05 '.' 0.1 ' ' 1
##
## Residual standard error: 1.031 on 280 degrees of freedom
## Multiple R-squared:  0.8749, Adjusted R-squared:  0.87
## F-statistic:   178 on 11 and 280 DF,  p-value: < 2.2e-16
```

The results look promising with an r^2 of 87%. There is probably not much that can be done to improve this model. With this step complete, we will assess how well the model does on the test data and create a plot in figure 10.2.

```
initialModel_pred<-predict(initialModel,test,type='response')
plot(initialModel_pred,test$Sales)
```

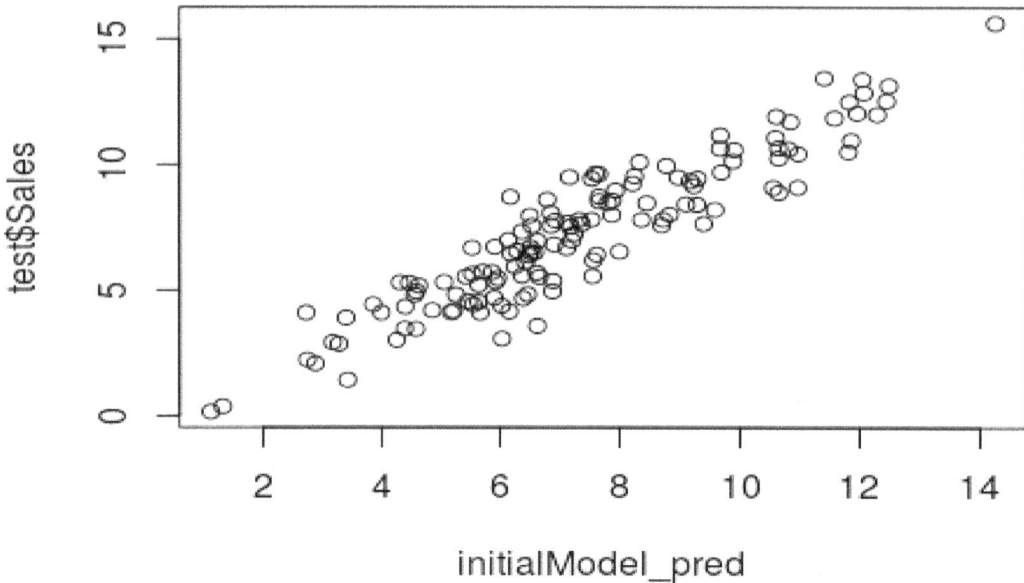

Figure 10.2: Initial model predictions vs. actual

You can see that the predicted values almost exclusively follow the actual values. Below we calculate the mean squared error and the rooted mean square error.

```
initial_model.resid<-initialModel_pred-test$Sales
#means squared error
mean(initial_model.resid^2)
## [1] 1.196961
#rooted mean squared error
sqrt(mean(initial_model.resid^2))
## [1] 1.094057
```

So far, we have three values that we can use for comparison with other models. The three values are.
- R-squared = .87
- Mean squared error = 1.19
- Rooted mean squared error = 1.09

The criteria for what is the best model will be based on these three terms.

Second Model

Below is the code for the second model. It's the same code as the first model except that we only used the following independent variables.
- Income
- CompPrice

- Advertising
- Age

Below is the code.

```
secondModel<-lm(Sales~CompPrice+Income+Advertising+Age,train)
summary(secondModel)

##
## Call:
## lm(formula = Sales ~ CompPrice + Income + Advertising + Age,
##     data = train)
##
## Residuals:
##     Min      1Q  Median      3Q     Max
## -8.2081 -1.7886 -0.1758  1.8184  8.0791
##
## Coefficients:
##              Estimate Std. Error t value Pr(>|t|)
## (Intercept)  8.127638   1.538197   5.284 2.53e-07 ***
## CompPrice   -0.001280   0.010345  -0.124   0.9016
## Income       0.013741   0.005505   2.496   0.0131 *
## Advertising  0.096548   0.023730   4.069 6.14e-05 ***
## Age         -0.039943   0.009546  -4.184 3.83e-05 ***
## ---
## Signif. codes:  0 '***' 0.001 '**' 0.01 '*' 0.05 '.' 0.1 ' ' 1
##
## Residual standard error: 2.624 on 282 degrees of freedom
## Multiple R-squared:  0.1272, Adjusted R-squared:  0.1149
## F-statistic: 10.28 on 4 and 282 DF,  p-value: 8.781e-08
```

This model is much worst. The r^2 is much lower at 12%. figure 10.3 is the plot of the predicted values. The code is as follows.

```
secondModel_pred<-predict(secondModel,test,type='response')
plot(secondModel_pred,test$Sales)
```

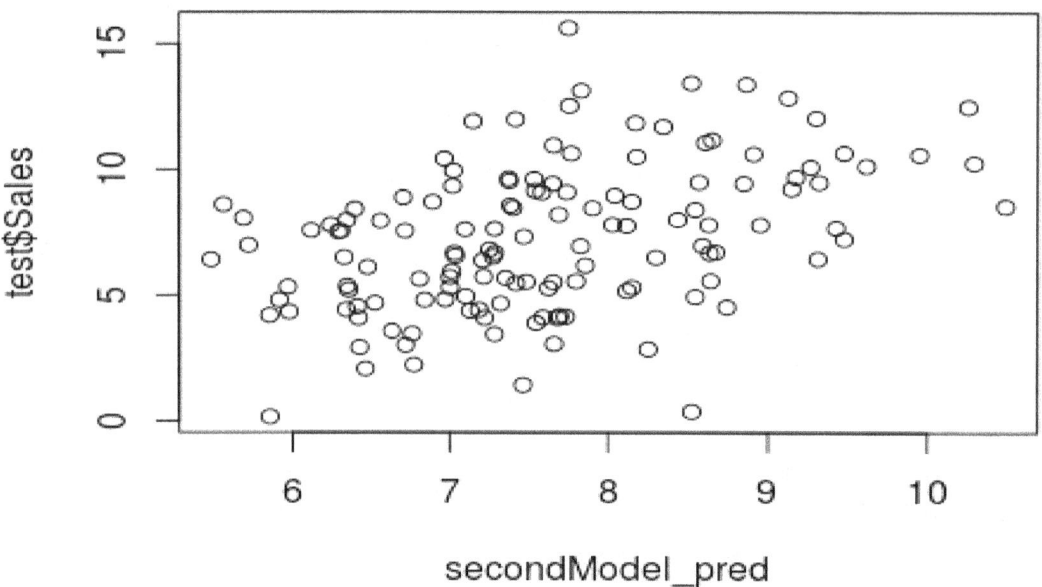

Figure 10.3: Predicted values of second model
    The predictions are much nosier. We will now calculate the error metrics.

```
second_model.resid<-secondModel_pred-test$Sales
#means squared error
mean(second_model.resid^2)
## [1] 7.024724
#rooted mean squared error
sqrt(mean(second_model.resid^2))
## [1] 2.65042
```

    We have already seen the difference in the r^2 values of the two models (.87 to .12). In addition, the mean squared error (7.02) and the rooted mean squared error (2.65) are much higher than our initial model 1.19 and 1.09. .Therefore we can conclude that the initial model is better based on these metrics of error and mode fit.

Cross-Validation
We can now see how our initial model performs in a different context using cross-validation. We will use k-fold cross-validation. This code is similar to chapter 9.
```
#Linear regression cross-validated
chck<-trainControl(method = "cv",number = 10)
customModel<- train(Sales~., data=train, method="lm",trControl=chck)
customModel
## Linear Regression
##
## 265 samples
```

```
## 10 predictors
##
## No pre-processing
## Resampling: Cross-Validated (10 fold)
## Summary of sample sizes: 240, 240, 238, 241, 237, 238, ...
## Resampling results:
##
## RMSE      Rsquared
## 1.002388  0.8771906
##
## Tuning parameter 'intercept' was held constant at a value of TRUE
##
predModel<-predict(customModel,test)
customModel.resid<-predModel-test$Sales
```

The RMSE is 1 with an r^2 of 0.87. These are highly similar values to what we got in the training set. This indicates that the model would perform in a similar manner with a different dataset.

Conclusion

Numeric models are normally judged by the amount of error that they contain. The actual value for the error is only useful when compare to another model developed in the same context.

Machine learning has grown tremendously over the years and many of the concepts of machine learning and data science are readily available to the typical data analyst. In this text, Darrin Thomas provides explanation and examples of the implementation of machine learning algorithms using R. Various concepts such as feature selection, classification, and numeric prediction are discussed.

Darrin Thomas, PhD, is Lecturer
at Asia-Pacific International
University located in Thailand

ISBN-13: **978-1546324195**
ISBN-10: **1546324194**

SuJinSoLa